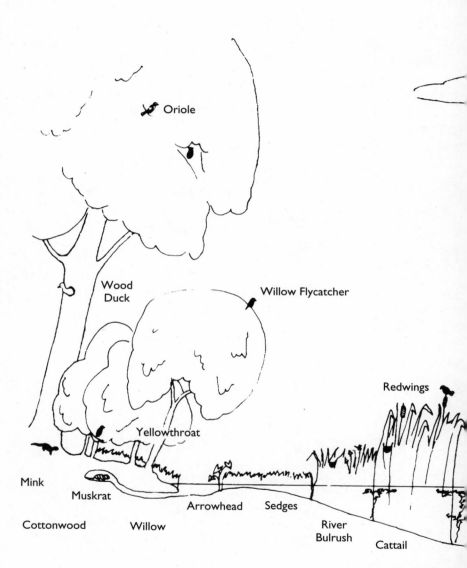

Oriole

Wood
Duck

Willow Flycatcher

Redwings

Yellowthroat

Mink

Muskrat

Arrowhead   Sedges

Cottonwood   Willow

River
Bulrush

Cattail

Bladderwort

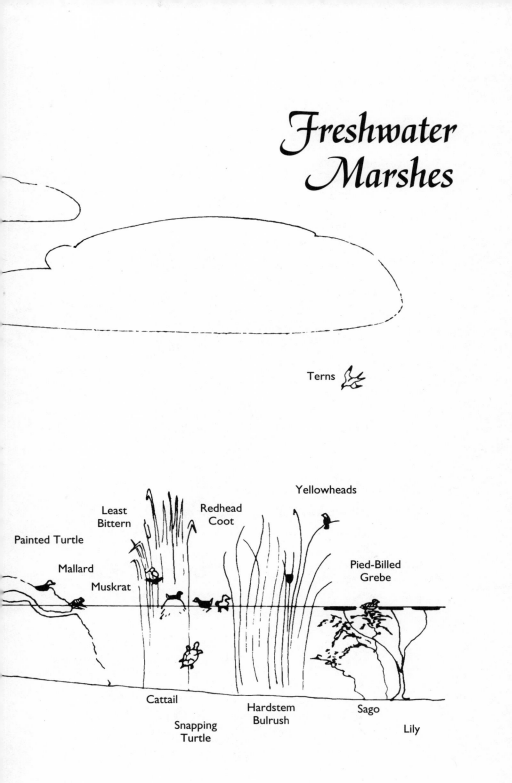

Freshwater Marshes

Terns

Yellowheads

Least
Bittern

Redhead
Coot

Painted Turtle

Pied-Billed
Grebe

Mallard

Muskrat

Cattail

Snapping
Turtle

Hardstem
Bulrush

Sago

Lily

WILDLIFE HABITATS
Milton W. Weller, Series Editor

VOLUME 3
Lowell W. Adams, *Urban Wildlife Habitats:*
A Landscape Perspective

VOLUME 2
Robert H. Chabreck, *Coastal Marshes:*
Ecology and Wildlife Management

VOLUME 1
Milton W. Weller, *Freshwater Marshes:*
Ecology and Wildlife Management, Third Edition

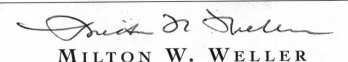

MILTON W. WELLER

Freshwater Marshes

Ecology and Wildlife Management

THIRD EDITION

UNIVERSITY OF MINNESOTA PRESS

MINNEAPOLIS / LONDON

Published by the University of Minnesota Press
2037 University Avenue Southeast, Minneapolis, MN 55455–3092
Printed in the United States of America on acid-free paper

Library of Congress Cataloging-in-Publication Data

Weller, Milton Webster.
    Freshwater marshes : ecology and wildlife management / Milton W. Weller. — 3rd ed.
        p.   cm. — (Wildlife habitats ; v. 1)
    Includes bibliographical references (p.   ) and index.
    ISBN 0-8166-2406-2 (hc : acid-free). — ISBN 0-8166-2407-0 (pb : acid-free)
        1. Marsh ecology.   2. Freshwater ecology.   3. Wildlife management.   I. Title.   II. Series.
    QH541.5.M3W34   1994
    574.5'26325—dc20   93-31760

The University of Minnesota is an equal-opportunity educator and employer.

The University of Minnesota Press
gratefully acknowledges assistance
provided for the publication of this volume
by the
John K. and Elsie Lampert Fesler Fund

To my wife, Doris

Contents

Preface

This little book on marshes and marsh wildlife is intended mainly to provide a general introduction for interested laypersons, students, and professionals in other fields. Experienced marsh specialists may find some of the viewpoints worth considering, but the text is not intended to be highly technical. Current interest in wetlands is at an all-time high, not only because of their rich wildlife diversity, but for their many other values.

Marshes have always been important to hunters and trappers, who in turn have been influential in their preservation. Fisherfolk also have recognized the values of marshes as spawning and rearing areas, and many are interested in their protection and management. Moreover, it has become obvious that marshes play important roles in assuring water quality, floodwater retention, enhancing water tables, erosion control, and soil and nutrient trapping, and are valuable in providing green space for esthetic and active recreation. Recent environmental protection laws have induced interest where it was not already present, and lawyers, sociologists, economists, consulting firms, developers, and land managers of all kinds are concerned about wetland evaluation and protection as they have never been before.

I have put together some general information about marshes and their wildlife, and the impact of society on them.

Anyone who tackles a task of this kind does so with certain biases. My major research and experience in North America have been with the birds, mammals, and plants of the diverse marshes in five areas: glacially formed pothole marshes in the prairies of Iowa, Manitoba, the Dakotas, and Minnesota; lakeshore marshes such as the Delta Marsh of Lake Manitoba; Great Basin marshes in Utah; the unique permafrost basins of the Alaskan

Arctic Coastal Plain; and coastal fresh and salt marshes in Texas. In addition to my personal observations, I have tried to broaden the scope of the book to current issues that are vital to understanding problems in the conservation of marshes. But because of my own interest and experience, the major emphasis is on the wildlife of marshes, the potential of and the procedures for management or restoration of marshes for wildlife, and the effects of natural forces and society on the functioning of this unique system.

Any brief treatment of a complex system suffers from problems of over-simplification. A marsh is a complex, living, dynamic ecological system (*ecosystem* of Odum 1971) with a characteristic physical *structure* that provides suitable habitat for members of a single *population* of a species, a diversity of substrates attractive to many species of plants and animals living together as a *community,* and a flow of nutrients (and therefore energy) from the nonliving materials in the system through the members of the community that allows the system to function and become self-perpetuating.

This book may produce mental conflict for some readers because it considers both the pristine and the manipulated, the economically tangible and the esoteric, the hunted and the nonhunted, and the conflict of wildlife with human beings as well as the benefits of wildlife to us. Management philosophies and approaches may be the most difficult to accept, but they must be viewed objectively in relation to the current state of the environment. I am the first to encourage managers to leave well enough alone, but marshes can not only profit from human manipulation, they may require it. The tools of management must be natural and effective, however, and the end product must be acceptable to society.

Attitudes toward marshes and marsh conservation vary dramatically. At one end of the scale is the drain, fill, and build-over syndrome; at the other end is the concept of preserving intact. Neither may be totally feasible in today's society, but perhaps the reader will be able to evaluate alternatives more objectively after reading this book. Solutions are not provided for all of the problems raised, nor are they likely to come easily, but a general knowledge and appreciation of the pros and cons will enable more effective decision making by humans, the present controllers of so many natural systems. If in the judgment of most of humankind, marshes are to be replaced by pastures and concrete, so it will be—but the world will not be the better for it.

Acknowledgments

Many people have influenced my thinking on marshes and have gone out of their way to help me to understand them better. Among the most noteworthy are William H. Elder (University of Missouri), Peter Ward and the late H. Albert Hochbaum (Delta Waterfowl Research Station), the late Paul L. Errington (Iowa State University), Leigh Fredrickson (University of Missouri), David Trauger (U.S. Fish and Wildlife Service), and David Voigts (Florida Power Corporation).

My research on North American marshes has been financed directly or indirectly by various organizations at various times in my career: the University of Missouri, the Delta Waterfowl Research Station, Iowa State University, the Iowa Conservation Commission, the U.S. Fish and Wildlife Service, the University of Minnesota, Texas A&M University, and the U.S. Environmental Protection Agency. I am indebted to many individuals in these organizations for their interest and assistance.

CHAPTER I
Introduction

Any low area that will hold water over soil, even temporarily, forms a suitable basin for the invasion of water-tolerant, rooted, soft-stemmed plants (hydrophytes) such as grasses, sedges, cattail, and bulrush. This semiaquatic, herbaceous plant community is a marsh, and it forms diverse habitats for many types of animals. The presence of standing water is variable, and even when standing water is not present, organic soils may hold sufficient water to promote germination or to sustain the growth of emergent hydrophytes. In some cases, hydrophytes may be missing for a period of time, but the soils are characteristic of those developed under water (hydric soil of Cowardin et al. 1979). These authors classify such marshes as "emergent wetlands," a designation that helps to separate them from other wetland types such as wooded swamps, where water-tolerant trees or shrubs are dominant, or moss-lichen bogs, where shorter and more fragile hydrophytes grow. Although we often speak of a marsh as a discrete entity, and this kind of marsh is the easier concept to identify and discuss, the term "marsh" also is applied to emergent marsh vegetation distributed as fringes along the shallow edges of lakes and seashores or as patches or strips along rivers.

Because of my own experience, and because the dynamics of such wetlands make them vulnerable habitats, most of the examples are from freshwater marshes of the glaciated prairies of the Midwest, which are typical of those at midlatitudes of the world where seasonal changes are dramatic. Emergent marshes may occur at any latitude, but they are poorly developed at very high latitudes (Arctic or Antarctic) because of the permafrost,

The prairie marsh adds water and subtle diversity of color and texture to the gently rolling grassland.

cold, and reduced productivity of such areas. In wet tropical regions, such marshes may exist if water levels allow, or may be replaced by shrub or tree swamps. It is the midlatitude marshes that are renowned for their rich productivity in terms of both the plant quantity (biomass) and the myriads of insects and waterbirds that have attracted our attention. Of all systems on earth, this type of marsh is one of the most productive in capturing and converting sunlight energy into living tissue. Although I focus here on inland systems, many of these concepts and principles are in evidence along the coasts where isolated freshwater marshes may be common and where even tidal marshes may be fresh or nearly fresh as runoff from the uplands is held back by the sea. Moreover, brackish or even saline marshes in estuaries and lagoons may show similar structural and biological patterns.

Each marsh is a complex community of living organisms interacting with their physical environment. The study of such communities is still incomplete, but the products of the research have broad implications for understanding natural systems in relation to human beings. Studies of a bog lake and of East Coast tidal marshes have been especially important in understanding energy flow and other processes in the ecological systems (Odum 1971). Work on coastal marshes has enhanced our knowledge of how marshes serve as nutrient traps that then enrich the adjacent ecosystem, the ocean (de la Cruz 1979, Teal and Teal 1969). Studies of freshwater marshes have focused more on the production of wildlife from such marshes (Bishop et al. 1979), the natural habitat features that influence wildlife use and diversity (Orians and Horn 1969), and the social systems within and between species (Orians 1961). Much of the work on both freshwater and marine systems has been observational rather than experimental, and only recently have replicated experiments become commonplace.

To facilitate a grasp of some terminology and various scientific perspectives, it may be useful to list the kinds of scientists who have studied marshes and how their diverse interests and approaches contribute to the understanding of this complex system. Formation of marshes usually involves landform, so geologists (specifically geomorphologists) concerned with glaciers, rivers, soil movement, and other dynamic physical forces clarify processes and actions by which basins are formed. Climatologists are concerned mainly with the weather-related phenomena such as rain, snow, and storms that influence marsh water levels. Hydrologists study the flow and accumulation of water in basins and matters pertaining to regulation of water levels. The study of water in such basins, and particularly the aquatic life of freshwater, is done by limnologists. Fortunately, a number of botanists interested in marsh and aquatic plants have written useful guidebooks to species identification (Fassett 1940, Hotchkiss 1970, and many other regional and state treatments), done detailed work on plant populations and nutrient dynamics in plants, and addressed issues of plant control (Riemer 1984). Work by ornithologists on the birds of marshes and by mammalogists on the mammals has provided data on the biology and behavior of various species, on species richness (i.e., the number of different species), and species diversity (species richness mathematically related to population size of each species). Few ichthyologists have studied the fishes of marshes, however, and more studies by herpetologists on the amphibians (frogs, salamanders, toads) and reptiles (snakes, alligators) of wetlands also are needed. Entomologists have done a great deal more work, but there is so much more to do, because insects are the most numerous group in terms of species richness and population size. Other invertebrate zoologists concentrate on mollusks, crustaceans, protozoans, and other groups of animals that lack backbones. Ecologists have concerned themselves mainly with the interrelationships of species, the evolution of communities of naturally associated species, and the processes that allow the ecosystem to function. Wildlife biologists have studied marshes mostly because the production of wildlife is dependent on the attributes of the marsh. Recently, restoration ecologists have focused on repairing the damage done by human perturbations to wetlands and other whole systems, and conservation biologists have worked toward identifying and preserving endangered ecosystems and species. This review is based mainly on the ideas and data of such scientists, and our knowledge of wetlands is advancing rapidly because of their fascination for marshes or their need to resolve some practical problem.

For readers who wish to explore some of these ideas in greater detail, or to interpret findings for themselves, literature cited in the text is listed alphabetically in the back of the book. Some suggestions for simple studies of marshes and their biota are given in Appendix A. For those interested in management or restoration, a more self-contained statement of principles and approaches is given in Appendix B. However, the steps are rarely as simple as they may sound, and expert advice is always recommended. Scientific names of plants and animals are listed in Appendix C, and a glossary of terms as used by wildlife biologists and wetland ecologists appears in Appendix D.

CHAPTER 2
Marsh Basins, Hydrology, and Diversity

Marshes may be formed in any basin that will hold water long enough for the germination and survival of semiaquatic plants. Basins probably hold water poorly when they are new, unless they are created in fine silt or clay that seals easily, but buildup of organic matter helps to fill the many pores. Eventually, basins collect groundwater, rainfall, snowmelt, or floodwater from watershed, river, or lake.

Basin Formation by Physical Forces

Wetland basins are landforms created by water movement of rivers or lakes, glacier or other ice action, tectonic action such as mountain building (Reid 1961), or even soil slippage (Weller 1972). Depending on local climatic conditions, extensive geographic regions may be characterized by one wetland type. The distribution of water-loving ducks is a good measure of both the distribution and the productivity of some important wetland zones (Fig. 1) (Kiel et al. 1972).

Glacial action, especially sheet glaciers of the Pleistocene epoch, formed the massive Prairie Pothole Region of the north-central United States and western prairie Canada. This area once covered some 270,000 square miles (700,000 square kilometers) (Kantrud, Krapu, and Swanson 1989) and probably was one of the richest wetland regions in the entire world because of the abundance of lakes, marshes, and smaller wetlands located in

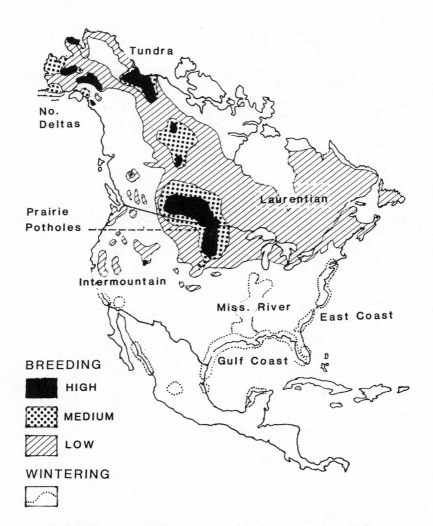

Figure 1. *The general pattern of duck distribution in North America during summer breeding (shading) and wintering (dotted line) serves as an index to the general distribution of various types of wetlands (Kiel et al. 1972). Major breeding areas are in the glacially formed Prairie Pothole Region, in the coastal tundra and riverine marshes of Alaska and Canada, and in the western intermountain and Great Basin marshes. Lesser numbers occur in the deep and sterile lakes of the LaurentianShield of Canada. Major wintering areas are in coastal lagoons and bays and estuaries of southern and eastern coasts, and along river oxbows and delta marshes of the Mississippi River, as well as in western basin marshes.*

rich soils with a warm summer climate. The areas of greatest pothole abundance are moraines of undulating glacial till. Once these sand and gravel basins were sealed with finer silts, water retention created suitable depths that induced growth of semiaquatic plants. Moraines of various types also influence stream and river flowages, with a resultant formation of major lakes in some cases and smaller, marshy basins in others. The ancient glacial Lake Agassiz of the north-central United States and southern Canada left undulations from old shorelines and underwater deposits. Extensive marshes and bogs have formed in these basins within the former lake bottom.

Intermountain depressions often are quite deep, and where water is abundant, lakes rather than marshes form. Ultimately, these depressions may be filled by eroded uplands to create more shallow, marshy basins, or they may dry out because of arid conditions. Water levels in intermountain valleys of the western United States often have snow as their chief source. Year-to-year fluctuation in snowfall in mountains quite some distance away may influence levels in the basin. Levels in lakes and marshes also may vary seasonally from the spring snowmelt period to late summer when water supply dwindles. In such arid western intermountain regions such as the Great Basin, major rivers pour into sumps, where no drainages exist and where water levels are decreased by evaporation and by plant transpiration. Such systems become highly alkaline or saline, such as Malheur Lake in Oregon (Duebbert 1969) or the Great Salt Lake of Utah (Christiansen and Low 1970). Concentrations of salts eventually may reach a stage where few plants or animals can adapt.

Beaches of gravel, sand, and soil deposited along the shores of major lakes by wave or ice action may create a marginal lagoon that traps water moving from the surrounding watershed. The resultant marsh differs little from other marshes except that periodic reflooding from the lake may change the chemical composition of the water and wash nutrients from the richer marsh system into the larger water body. The vast marshes of this type along the Great Lakes and along the huge prairie lakes of Manitoba (such as Lakes Manitoba, Winnipeg, and Winnipegosis) not only are major waterfowl producers (Hochbaum 1944), but strongly influence the quality of fishery resources in those lakes. The productivity of such areas is influenced by the degree and rate of nutrient exchange between the systems. A comparable but even more dynamic system occurs in coastal marshes along oceans, where tides may flush the vegetation daily and enrich adjacent shallow marine areas (Chabreck 1988).

Rivers form marshes indirectly through their meanders as they cut new channels and abandon old courses. Oxbow lakes result, and siltation eventually produces oxbow marshes that tend to become progressively more shallow and less lakelike. Because of their drainage of large areas, such marshes may be extremely rich, but the character of their plant and animal species often is influenced by annual flooding and resultant siltation. Although such flooding resupplies nutrients annually, species of perennial plants that will not tolerate periodic flooding and drying are reduced or eliminated. Plants that are well adapted to this unstable system, such as annuals or perennials that establish easily by seed, are highly productive and become dominants. As semiaquatic plants stressed by such silt die out, shrubs and trees may invade. The plant production resulting from these areas is washed seaward whenever floods fill the oxbow, and marsh and river are contiguous.

Marshes also are found where major streams form deltas or spread out as braided streams on extensive flood plains. Marshland of the Mississippi River Delta represents this type of marsh (Gosselink 1984), and similar extensive riverine marshes occur elsewhere in the world. In Iraq, the deltaic marsh system of the Tigris and Euphrates rivers covers thousands of square miles and is the home of an entire human culture that lives in and depends on the marsh for food and housing (Maxwell 1957).

In arctic and alpine areas, frost action segregates rock and soil particles of various sizes and shifts them in such a way that "polygonal earth" patterns are formed. Those with low centers become basins that capture snowmelt (Bergman et al. 1977). In the arctic coastal tundra, characterized by permanently frozen (permafrost) shallow organic soils, sun heating (insolation) of snowmelt water results in the formation of basins by the thawing of the ice substrate. Further warming of the water in the pool thaws the substrate until an equilibrium is reached between the temperature of the sun-heated water and the chill of the permafrost basin underneath. Emergent plants survive only in the more shallow basins because they often dry out early in the season, allowing invasion by a few species of sedges and grasses.

Biological Influences on Wetland Structure

The major biological influence in formation of marshes is the dense growth of diverse plants that capture sunlight and build plant biomass, slow water movement, seal basins, induce settling of particulate matter from the water, and protect extensive areas from wind action. These actions generally reduce basin depth by means of deposition of peaty organic materials. Depending on the slope of the land, water-retaining mechanisms of plants aid in building their own wetland, such as occurs in the extensive sedge meadows and blanket bogs of northern forest areas.

The major animal builder of wetlands is the beaver, but in most cases, fairly deep and open "ponds" are formed rather than densely vegetated marshes (Beard 1953). Nonetheless, they provide similar kinds of wetland habitat in an otherwise streamlike environment. Often these ponds have rather abrupt shores, but spikerush, sedges, and, occasionally, broadleaf cattail are found in the upper reaches. Water lilies and other pad plants are common in the deeper water.

Muskrats, nutria, and snow geese feed on rootstocks, which results in opening and deepening of marshes (Glazener 1946, Lynch et al. 1947). Through the action of carp and muskrats (in combination with waves and ice), large areas of plants may float to the surface, carrying soil with them in the rootstocks. Such actions aid in deepening marsh basins, but also create turbid water. Depressions of various sizes are created by crayfish, muskrats, alligators, seals, elephants, and other animals, in various geographic areas of the world. These deepen small areas of a marsh and may strongly influence the rate at which the bottom peat layer builds up and where plants are distributed. These animals are important influences on wetlands though not the causative agent forming major aquatic systems.

Hydrology

One usually assumes that water in marshes is a product of direct rainfall, snowfall, or runoff from adjacent slopes, or at most the overflow of rivers in riverine marshes. Although hydrologists have only scratched the surface in the study of wetlands (Winter 1989), they have found diverse patterns. Some emergent marshes may have water levels that are not influenced by the deep ground water table but are in "sealed" or "perched" basins above the level of that table, which in part explains their highly variable water levels. But extensive wetland areas such as the Prairie Pothole Region may have water levels that reflect the water table and contribute to it. Especially when marsh water levels are high, seepage from marshes into the groundwater occurs mostly at the margins. For this reason, seepage losses are greatest where wetland shoreline is largest in proportion to the water volume (Millar 1971). In several areas where the seepage into the ground level has been calculated, it tends to be less than 20% of the total water volume of the wetland each year, but that amount can be a major contribution to the watershed and to water quality (Allred et al. 1971, Eisenlohr et al. 1972). Subsurface water flow may intersect and even interconnect potholes (Sloan 1972). This water table can slope into or away from a pothole, creating a "discharge" if it slopes into a pond, and a draining "recharge" if it slopes away from the pond. In some situations, this creates a flow through of water, but much can happen to water while in the wetland. Considerable water is lost owing to evaporation, and plant transpiration can be even more significant. In fact, transpiration loss often exceeds that due to evaporation—and is greater with tall than with short emergent plants. Because both occur during the summer, the combined loss of water through evaporation and transpiration was two and a half feet (seventy-six centimeters) in some North Dakota potholes, far exceeding annual rainfall of one foot (thirty centimeters) (Eisenlohr et al. 1972). Thus, runoff was essential for maintenance of the water level, and snowfall, early spring rains, and carryover were essential for maintenance of levels under dry conditions. A slight modification of these water supplies shifts the prairie potholes to "wet" or "dry" cycles in a very short time. Those areas with water high in dissolved salts that also have high summer evaporation become alkaline (sulfate) or saline (chloride) wetlands, with a corresponding change in vegetation and animal life (Stewart and Kantrud 1972). Variation in flow-through rates gradually modifies the water chemistry of the area and, eventually, the emergent vegetation, so that flushing with fresh water is a regular management practice in western (Christiansen and Low 1970) and coastal saline marshes (Chabreck 1988).

It should become obvious from these generalizations that wetlands are vital to continental water resources. They filter water as it moves to streams and influence the rate of flow (especially during storms); they also may act as storage basins during floods (Moore and Larson 1979). At some times of year (especially spring), they may increase the flow rate of streams that draw their water from watersheds rich in wetlands, whereas such basins may hold water in dry periods (late summer) and reduce stream flow (Barney 1980). Wetlands have a strong influence on water quality since sediments (as much as 80%) and heavy metals are deposited, and nitrogen, phosphorus, and other nutrients are extracted and modified by cycling within the marsh system (Kadlec and Kadlec 1979).

Some bog wetlands actually may draw water from lower levels by capillary action in the organic substrate. In a few cases, wetland plant associa-

This large open wetland in North Dakota is dissected by a county road. Drainage into the wetland is increased by the road surface and the ditch formation. Salts from winter ice removal undoubtedly contaminate the water.

tions form on hillsides where groundwater discharges, so that moss, sedges, and cattail can be seen in patches on otherwise dry road cuts or hillsides.

Rainfall Patterns, Hydroperiods, and Wetland Types

Much of the emphasis of this book is based on research and observations in the shallow temporary to the deeper semipermanent prairie potholes of the Upper Midwest. Thus, the dynamics of wetland water regimes, temperature ranges, and other environmental influences are characteristic of that region. Wetlands elsewhere have the same responses to water and may have many of the same plants and animals, but regional differences are many, and they usually are products of different water regimes produced by amount and seasonality of rainfall or snowmelt. This results in different hydroperiods (duration of water presence), seasonality (spring versus winter or summer), and depth. Thus, some regions are characterized by deep, lakelike water bodies where emergent marsh vegetation is rare. By comparison, more shallow basins tend to be densely vegetated. To provide some feeling for the mean water regimes that create or maintain these various wetland types, and to aid in appreciation of the forces that drive the plant community and influence how animals use them, a schematic model is shown in Figure 2.

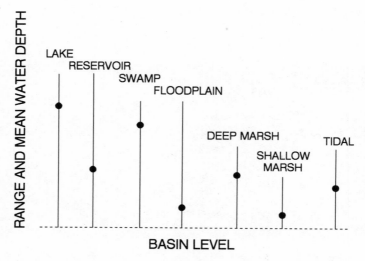

Figure 2. *A schematic model shows the range of water depths and mean annual fluctuations in various types of wetlands to contrast water regimes in typical prairie marshes classed as deep marsh or shallow marsh. See text for details.*

Lakes and their constructed equivalent, reservoirs, are characterized by long hydroperiods and deep water that limit the presence of emergent plants. Submergent plants may form submerged aquatic beds in more shallow areas of three feet (ninety-one centimeters) to fifteen feet (4.6 meters), depending on the suitability of substrates, wave action, and water clarity. Reservoirs typically are manipulated to store water or to release flood- or irrigation water, and thus their water levels vary more than those of lakes. Accordingly, I have indicated that their annual mean tends to be lower than that of natural lakes, but this would vary with area and water regime (Fig. 2).

Swamps are wetlands dominated by woody trees or shrubs, which tend to have long and fairly deep water that prevents invasion of emergent plants. They often are in basins carved by the river, and thus they hold water longer than do the bottomlands that are temporarily flooded to great depths but where the water runs off easily. Although water depth in forested bottomland wetlands may be very deep for a shorter period as a result of river flooding, such bottomlands typically are flooded during the trees' dormant period. If flooded very long during the growing season, many species would die and the area eventually would become either marshlike (as sunlight entered a formerly shaded area) or dominated by still more water-tolerant trees (Weller 1989a).

Typical prairie wetlands are indicated in Figure 2 as "shallow marsh" and "deep marsh." Like swamps, "deep" marshes also have long hydroperiods but vary in water presence and depth by season and between years. Because marshes are more shallow than lakes or even swamps, they are ideal for emergent plants in water less than thirty-nine inches (one meter) deep. There is no clear-cut demarcation between deep and shallow marshes or the still more temporary moist-soil and meadow areas. This is a continuum from deeper and therefore more permanent to more shallow and less permanent, much like the range of water depths from edge to center of any basin wetland. Deep marshes stay wetter longer on the average, but still dry out periodically—usually not every year, but once every several years—and at that time are populated by the more water-tolerant but herbaceous, perennial emergent plants that spread by propagative growth but that can also seed. Shallow wetlands are less likely to have the tall and persistent bulrushes and cattails that we see in deeper marshes simply because those plant species are better adapted to grow and to survive in relatively deeper water. The more shallow areas are dominated by a mix of perennials and annual plants like water-tolerant grasses, sedges, arrowheads, and smartweeds. Such areas grade into seasonally flooded, meadow-like areas that dry early and are covered by herbaceous vegetation that are mostly annuals and low-profile emergents such as spikerush, rushes, and water-tolerant grasses.

Tidal areas have quite different water regimes because they have regular hydroperiods dictated by ocean tides (see Chabreck 1988). Typically this event occurs twice every twenty-four hours, but it varies from one to three times a day in various areas at different times of the year. Hence, the annual mean is a figure more nearly midway between high and low extremes than found in the more variable inland wetlands (Fig. 2). Although one might assume that such tidal patterns are not linked to freshwater marshes, that is not always true. Especially on the East Coast, but occasionally on the Gulf Coast as well, streams entering the brackish estuary are backed up by the tide. Thus, shallow and slow-moving streams may support marsh vegetation and are fresh in their upper reaches most if not all of the time, yet they still have tidal action.

A small temporary wetland in early summer was plowed nearly to the edge. Although too little upland remains for nesting, pairs of ducks may find isolation and food there in spring, and shorebirds find such wetlands useful during spring or fall migration.

Wetland Diversity and Complexes

Because of the dominance of physical–geological forces responsible for the formation of most wetland basins, multiple wetlands usually are formed over a large area during a geologic period. Often, these same forces created wetlands of different depths, sizes, and intermarsh spacing in a general area. The association of several small temporary or seasonal wetlands around a large body of water provides the greatest diversity or heterogeneity in a complex or cluster that can serve a variety of birds and other wildlife (Flake 1979, Patterson 1976, Weller 1979). The diversity produced by clusters of wetlands has great importance to habitat use by various species of wildlife. Thus, we should attempt to maintain this diversity of landforms and vegetation types in as natural state as possible and with a balance characteristic of pristine times.

Classification

Wildlife biologists and other wetland scientists have devised a number of systems for use in surveying marshes and their wildlife (Bergman et al. 1977, Cowardin and Johnson 1973, Golet and Larson 1974, Jeglum et al. 1974, Martin et al. 1953, Millar 1976). Any classification system involves artificially dividing what is really a continuum; hence, some wetlands will seem to fall between any two categories and may be judged differently by different classifiers. In most cases, this apparent discrepancy is probably not serious, and any one person will be fairly consistent in using a system. As we shall see, these types of wetlands differ in their attractiveness to various groups of birds and other wildlife.

Several systems currently are in use in the Prairie Pothole Region, and may serve investigators in different ways. Shaw and Fredine (1956) (Table 1) categorized marsh and other wetland areas of the United States in a system that long was the legal basis for decisions on wetlands, known colloquially as "Circular 39" of the U.S. Fish and Wildlife Service. Water

Diversity of wetland types and sizes is especially attractive to many marsh birds and probably is required to attract species with large home ranges. Dewey's Pasture Wildlife Management Area and National Historic Landmark near Ruthven, Iowa, is an "island" of natural wetlands in the midst of intensively farmed areas, consisting of about four hundred acres, probably sufficient here in size because of the attractiveness of the large, shallow, food-rich lakes adjacent to it. Photo by Tom Rothe.

Table 1. A partial comparison of two wetland classification systems

Shaw and Fredine 1956 types	Stewart and Kantrud 1971 classes	Major identifying vegetation
1. Seasonally flooded basins or flats	I. Ephemeral ponds	Wet-prairie grasses, annual weeds
2. Inland fresh meadows	II. Temporary ponds	Meadow sedges, rushes, grasses, broadleafs
3. Inland shallow fresh marshes	III. Seasonal ponds and lakes	Water-loving grasses and sedges, smartweeds, burreed
4. Inland deep fresh marshes	IV. Semipermanent ponds and lakes	Cattail, hardstem bulrush, submergent pondweeds
5. Inland open fresh water	V. Permanent ponds and lakes	Type IV emergents along the shorelines
10. Inland saline marshes	VI. Alkali ponds and lakes	Alkali bulrush, wigeongrass

depths and hydroperiod were emphasized in this work as influences on the vegetation that typically develops under those water regimes. Hence, a vegetation type was inferred that we commonly use in referring to plant zones: deep fresh marsh, shallow fresh marsh, and so forth (Table 1). Stewart and Kantrud (1971) (Table 1), working mainly in the North Dakota glaciated prairie wetlands, emphasized the relative permanence of the wetlands in the class names, but made more extensive analysis of the vegetation cover type by depth zone, as well as the changes in vegetation characteristic of the area and the wetland class. Table 1 shows both systems, which are reasonably comparable except for inland saline or alkaline areas, which we will not treat here in detail.

Because of the interest and involvement of various agencies and interest groups, more refined wetland classification systems have become essential for application to various wetland types and conditions. The current standard classification system devised for mapping of all kinds of wetlands (Cowardin et al. 1979) is hierarchical (a graded series from general to identifying specifics). First, the wetland is sorted by general characteristics using established scientific terms that avoid the confusion of local common names, and is given a designation called a System: Marine (the intertidal zone at the edge of the sea), Lacustrine (shallow fresh lakes), Estuarine (brackish and salty coastal wetlands), Palustrine (literally "marshlike" but it includes shrub and forested swamps as well), and Riverine (wetlands within river channels rather than the Palustrine oxbows along them). Within all of these Systems (except Palustrine) are Subsystems (such as subtidal or littoral), and all have a third (lesser) hierarchical level termed class. The Classes include prominent features that aid in identification such as emergents, aquatic (i.e., submergent) plant bed, and rock bottom. Subclasses then point out characteristics shared by many wetlands in the class such as Persistent (dead plant material that survives into the next season) or Nonpersistent (decomposes rapidly). A Dominance type may be used to list characteristic plant species, and other special modifiers denote water regimes, chemistry, human construction, or modification (Cowardin et al. 1979). An example of a freshwater marsh using this hierarchical system is as follows:

System	= Palustrine (marshlike)
Class	= Emergent Wetland (vs. Forested, Shrub)
Subclass	= Persistent (emergent plants that persist until the next year as dead plant material vs. nonpersistents that decompose in a few months)
Dominance Type	= e.g., Cattail or Cattail/Bulrush

To date, classification systems have been used more for censusing wetlands than for wildlife surveys, but several waterfowl and waterbird surveys have been made with the use of this system. These studies show associations of certain bird groups with certain wetland types (Bergman et al. 1977, Stewart and Kantrud 1973). Such data not only help us to assess what is present, but also allow us to predict what would happen if we intentionally were to change wetlands from one type to another.

CHAPTER 3
Marsh Substrate and Vegetation Structure

Marshes, like all plant communities, have certain physical features such as vegetation structure or life form that provide identifying characteristics (Beecher 1942). In addition to water, the main visual clues are the emergent plants of various heights that form the perimeter and sometimes the center of a marsh. These are taller and more robust than most prairie grasses, but neither as tall nor as woody as shrubs and trees. As we examine the marsh structure from shore to center, we find other more aquatic plants, as well as water and basin characteristics that mechanically influence how animals are attracted to and use the community.

Substrate

The composition of the basin, usually of fine clays and silts enriched by organic matter, but occasionally of sand or gravel, dramatically influences the available nutrients, the rate of pioneering of plants on that substrate, the survival of plants, and the durability of the bottom itself. The character of this bottom also determines the nature of habitats available for invertebrates, fish (especially for spawning), and the influence of wave action on rooted plants. These generalizations will gain significance in later discussions of other structural components of the system as habitats for animals.

Islands of cattail dot an impounded area of the Roseau River Wildlife Management Area, Minnesota. These islands are ideal areas for nesting ducks and coots. Dense beds of the submergent water milfoil are visible at the surface.

Water–Depth Influences on Plants

The distribution of plants of various life forms determines a major structural feature of a marsh. Water depth is perhaps the dominant physical factor influencing the kind of adaptations required of the plant if it is to establish, live, and reproduce on a site (Ovington and Pearsall 1956, Robel 1961, 1962). The life-history strategy of the plant is a product of evolutionary adaptations that include reproductive system (seed or vegetative propagation), leaf form, stem or stalk form, root structure, water tolerance, salt tolerance, and other physiological factors (Riemer 1984, Sculthorpe 1967).

A few examples will provide some patterns among common species. The cattails constitute a well-known plant genus that dominates wetlands around the world. Physical, biological, and physiological responses to water are so varied among species that the resultant, localized forms have confounded the scientific classification of cattails, and specialists disagree on whether some forms are hybrids or valid species. But more important for animal residents, cattails of various species or populations grow in waters of various depths, creating food supplies for herbivorous insects and muskrats, structural support for bird nests or bird roosting, and building

Floating cattail was uprooted by flooding of buoyant rootstocks and possibly by muskrat "rooting." Hardstem bulrush is in deeper water behind.

materials for nests of birds and semiaquatic mammals (Geis 1979, Linde et al. 1976, Weller 1975a).

But cattails aren't found everywhere; they have physiological and physical limits—as do all organisms. Field observation and greenhouse experiments show that some forms of cattail germinate and grow better in deep water, that they produce more underwater tubers (growth shoots) per plant, and that plant density (plants per unit area) is greater at optimal than maximal water depths (Grace 1989, Weller 1975a). The typical pattern in the Midwest is that broadleaf cattail is found in shallow water at marsh edges or in more shallow, temporary wetland types, whereas narrowleaf cattail tends to be in deeper water. A hybrid between these two is the most water tolerant of all, sometimes growing in three feet (ninety-one centimeters) of water, as also is true of southern cattail. Studies of the physiology of cattails have shown different efficiencies of growth and nutrient storage at different depths (Grace 1989). Moreover, the physical influences of wave action and substrates suitable to hold the plants (Weisner 1991) also are factors, as are tuber feeding by muskrats and food searching by fish such as carp, which can cause flotation of the buoyant rootstocks (Weller and Fredrickson 1974). The edge between a cattail bed and open water becomes a tension zone. The plant edge—water interface or ecotone shifts back and forth from season to season with changes in water depth, wind action, and animal density.

At the shoreward edge of the wetland, it is probably the availability and constancy of moisture, competition with better-adapted shallow-marsh plants, and perhaps nutrient dynamics that influence the success of cattail and hence their extent in the uplands edge.

Arrowhead is another widely recognized species of the marsh-edge zone typically inundated early in the spring. In these sites, most species form broad and firm lanceolate or arrowhead-shaped leaves. If, during drought, they become established in the marsh center and later are flooded, rather than arrowhead-shaped leaves, they develop soft and filiform leaves that make identification challenging for experts trying to distinguish them from other normally underwater species like wild celery. Often, as they reach the surface, their leaves broaden, the stems become more rigid, and they flower. Ultimately, with continued high water, they die out.

These examples demonstrate that plants found in a wetland are adaptable but also have limits, and that the life-history strategies of individual species collectively influence the kind of community of plants present. This plant community, in turn, dictates what animals are present.

Plant Life Forms

As a product of years of evolutionary adaptation, various groups of plants have evolved different strategies for different water depths, and they can be classified into several major groups. Emergents are plants that grow with their roots and often bases in wet soil or water part or all of their life. Examples, given from shallow to deeper water, are rice cutgrass, whitetop grass, sedges (softstem bulrush, river bulrush and other "three-squares"), cattail, and hardstem bulrush. The more robust species that stand into the next year have been termed persistent emergents, whereas those that deteriorate rapidly are considered nonpersistent (Cowardin et al. 1979). Floating-leaf plants are those rooted in deeper water that tend to send up broad, floating leaves to the surface where photosynthesis takes place. Nutrients move between leaves and massive tubers through flexible and slender stems that may be five or six feet (1.5 to 1.8 meters) long. Thus,

Submergent bladderwort, widespread throughout temperate and subtropical North America, has small animal traps that supplement its nutrient intake. The small floating plants are lesser duckweed, and the starlike plants near or under the surface are star duckweed.

such plants can grow in much deeper water than can most emergents, filling in otherwise open pools, and they survive well with fluctuating water levels or high turbidity. An example is yellow water lily. Submergent plants generally are rooted but have their stems and leaves mostly if not entirely underwater (Swindale and Curtis 1957). Characterized by fine, complex, and compound leaves growing in clusters, they seem to be efficient at gathering the vital light in even murky water. One group, bladderwort, has combined green leaves with small bladderlike animal traps in which the plants catch and utilize minute crustaceans and protozoans. Numerous species of submergents flower at the water's surface and bring color to the marsh—yellow for bladderwort, white or yellow for watercrowfoot. Some, such as sago pondweed, also form seed heads on the surface and are food for insects and ducks. Other examples of submergents are water milfoil, and wigeongrass in alkaline or brackish waters. Floating plants are not rooted to the substrate but do have dangling roots that derive nutrients from the water. They seem to be fairly small in the northern latitudes, where, due to freeze-up, they are annuals; and they are larger in warmer climates, where they may be less influenced by seasonality. Such free-floating plants are strongly influenced in distribution by the wind, tending either to windrow or to remain between protecting emergents, except in tiny wetlands where wind is less influential. Examples are lesser duckweed and, in the South, introduced water-hyacinth. Some floating plants, such as star duckweed, are found below the surface in the water column, and drift with minute water movements induced by animal activity and probably differential heating.

The combination of life forms and species in a particular wetland constitutes the living, dynamic, interactive plant community. Sometimes such plant communities last for many years if they are perennials and the water conditions are ideal. Sometimes they change from season to season, especially at the edge where annual plants dominate and water levels are most variable. The question of why each plant exists in a certain marsh is outside the scope of this book, but some influences will be discussed later under marsh habitat dynamics. Our immediate concern is how these forms in themselves survive and influence other organisms in the community.

Marsh Islands, Edge, and Layers That Influence Wildlife Use

Because marshes typically are formed in basins, they are surrounded by uplands with diverse kinds of vegetation. These wetlands are thus habitats nearly as different as islands in the ocean; indeed, there are some advantages in viewing marshes as islands, since concepts of island biology have been fruitful approaches to understanding how animals and plants reach isolated habitats and are affected by the size, placement, and character of these habitat units (Weller 1979). Marsh basins become concentration areas of nutrient material in the form of decaying organic matter and minerals that wash in. The productivity is high, and we can expect dense concentrations of living organisms that use the nutrients, as well as others that feed on those organisms. Concentrations of birds generally attract predators, but terrestrial predators are inhibited by the presence of water. Only a few specialized predators (mink, otter, and perhaps raccoon) regularly swim and live in or near water, even rearing their young there. Vegetation in the central marsh often is protected by water, and bird species—such as blackbirds, egrets, herons, and ibises—not only concentrate there but also tolerate a short near-neighbor distance, which allows them to form dense colonies and realize a high reproductive success in the absence of predators. The same phenomenon occurs on small islands in lakes where gulls, terns, and ducks have higher-than-average egg success in a predator-free environment, and where they and their offspring return to breed regularly (Koskimies 1957, Vermeer 1968, Lokemoen and Woodward 1992).

Rock or soil islands, or islands of bulrushes, reeds, or trees in marshes, and even muskrat lodges have an especially dramatic effect on species richness of birds and how birds place nests or obtain food. Islands, like irregularly shaped marshes, create additional edge, and there is good evidence that marsh-upland edges or cover-water edges increase both population density of some bird species and bird species richness (Beecher 1942). The absence of emergent vegetation in a marsh causes an opening, or pool, often in the center but possible anywhere in slightly deeper water. As with islands of vegetation, the size, abundance, and distribution of such pools modify the amount of cover-water edge and the access to cover, and seem to be important in creating more diverse habitats for birds.

The interface of marsh edge and water is equally important for the more aquatic species such as fish. Even fish that live in open water may spawn and grow in the protection and shallows of marshy areas. More aquatic, strong-swimming birds such as pied-billed grebes and diving ducks may feed on fish and invertebrates in open ponds or lake, but nest and rear their young in the emergent plants of marshes or marsh edge.

Another approach to understanding habitat influences on animal distribution is the concept of specialization of some animals for specialized habitats. In forests, layers or strata of vegetation are important influences on the numbers of species present, since some birds are adapted to treetops, some to trunks, and some to understory vegetation or ground cover. Generally speaking, the more strata, the more places for different species to feed or nest. Marsh-nesting birds also use several different layers: water level (pied-billed grebes), low vegetation near water level (coots, several ducks, rails), robust emergents like cattail (blackbirds, egrets), and marsh-edge trees (northern orioles, wood ducks) (Weller and Spatcher 1965).

CHAPTER 4
The Marsh as a System

Because all organisms require food for their survival, growth, and reproduction, many behavioral and physiological activities of plants and animals are devoted to food getting. A study of a marsh community focusing on the food interrelationships of its members will reveal much about the efficiency and the total energy of the system.

Various organisms in the wetland community fulfill different roles or niches, and the system can function well only when all components are present and effective. Plants fill the role of primary producer, converting water and carbon dioxide into carbohydrates through the energy of sunlight and the action of chlorophyll. A few marsh and bog plants are carnivorous, but their greenery indicates that trapped animals fill only part of their food requirement. But plants also create a physical environment, trapping heat, reducing wind, stabilizing soil, and providing substrates as well as food for animals. All animals are dependent on plant producers; thus, we cannot discuss animals alone in any ecological sense. Those animals using only plant food directly are termed herbivores, and are the primary consumers in the flow of energy through the system (measured as calories). Some animals feed on the herbivores and are therefore carnivores, in a secondary consumer role; those higher carnivores that feed on secondary consumers may be termed tertiary consumers. Some animals, omnivores, use both plants and animals as food. An often unappreciated group of organisms (the detritivores) assist in the breakdown of organic material resulting from growth of plants such as emergents. Some of these are shredders like muskrats that eat some small portion of the total produc-

tion of emergent plants but shred still more in cutting for lodges and select-
ing food, thereby creating smaller pieces. In stream systems, there may be
five or six insect or other invertebrate groups that cut and shred various
sizes of plant debris until it can be used by smaller and smaller organisms.
Physical action like water and wind also helps to produce detritus from
larger plants or animals. Eventually, microscopic decomposer organisms
like bacteria and fungi act on the organic matter, breaking it down to
organic compounds usable again by plants. Ultimately, nitrogen and phosphorus, vital to protein synthesis and energy metabolism, respectively, and
other important elements must be available or the productivity of the
entire ecological system suffers.

Food Chains and Webs

Specialization for foods in a marsh habitat is no different from that in
other communities, except that animals adapted to marshes have to be able
to move in and around water and take food from water as well as near it.
Terrestrial herbivores readapted to the marsh by virtue of webbed feet
(birds and mammals), heavier down (birds), or water-tolerant fur (mammals) would not be likely to change their food habits greatly. Thus, the
muskrat is a member of the subfamily of field mice (Microtinae), but it has
webbed feet and a flattened tail useful in swimming, and water-resistant fur
groomed regularly to ensure dryness and warmth (Errington 1963). But it
is merely a well-adapted plant feeder that utilizes marsh emergents (cattail,
bulrush, arrowhead) rather than terrestrial grasses. It builds a nest of shredded vegetation like other field mice but may locate it inside a massive lodge
of cut marsh plants, usually built in the fall before winter freeze-up
(Errington 1957). This herbivore plays the same role as a herbivorous
insect but on a larger scale. This is an example of the simplest and most
direct of all food chains, moving nutrients from the plant that produces the
food to the tissue of the primary consumer, with some loss in tissue conversion and in metabolism by a mobile organism (Buchsbaum and
Buchsbaum 1957, Odum 1971). Among truly aquatic forms, the consumption of algae by small free-swimming zooplankton (protozoans and
crustaceans), which in turn are eaten by small minnows, which subsequently may be eaten by larger fish, which then are eaten by a heron, provides a classic example of a longer food chain involving secondary and
tertiary consumers.

Most food chains are not tightly structured because animals must be

A muskrat feeds on cattail alongside its lodge. Lodges may contain nests and families, or groups of five to eight individuals of various ages.

opportunistic feeders, using what is available at a given time. As a result, food interrelationships are so complex they are spoken of as a food web. In certain situations, animals that are normally prey may become predators. Omnivores have especially great flexibility, which may explain why this group is so numerous. Muskrats, normally herbivorous, may feed opportunistically on clams (whether by nutrient deficiency, local habit, or design); blackbirds may feed their young on invertebrates from the marsh or fly some distance to gather terrestrial insects and even grain where they can. Herons, which normally feed on fish and amphibians, may eat other birds' eggs, terrestrial snakes or small mammals, and diverse other foods. Herons' basic foraging behavior and predatory habits are generally unchanged, regardless of their foods, but they adapt to food availability readily. Little blue herons in the South sometimes follow mergansers and pied-billed grebes that are feeding on fish, the herons apparently catching some fish attempting to escape from the underwater divers (Emlen and Ambrose 1970). This effective adaptation is accomplished by "association" rather than "calculation," but it works!

These organisms are part of the complex system of nutrient movement through the community, so obviously the supply of various nutrients controls the relative "richness" of the system.

Nutrients

Water areas may be classified in relative terms of nutrient content and productivity. Most marshes are rich (eutrophic) because of the organic buildup; lakes, especially those with rock or gravel basins, often are more sterile (oligotrophic) (Odum 1971). Such nutrient levels are reflected in the richness of plants, invertebrates, and even waterbirds (Reichholf 1976, Utschick 1976, Weller 1972). Where cold temperatures slow the chemical processes in plants themselves, production is naturally low at all levels, and evolution of the system is slower. Therefore, time, temperature, seasonality of growing seasons, and other environmental factors influence the buildup of nutrients. In some antarctic ponds, it seems certain that excrements from birds and seals—the foods of which come from the sea—are the major source of nutrients for unicellular aquatic plants responsible for energy trapping and subsequent food chains (Weller 1975b). Concentrations of birds swimming and feeding in wetlands also may alter water quality and characteristics (Ganning and Wulff 1969, Have 1973, Manney et al. 1975), and farm drainages from livestock holding areas quickly enrich wetlands and modify the nature of the vegetation there.

Wetlands gain nutrients from several sources: Inflowing water and silt enriched with phosphorus, nitrogen, and other potential nutrients adds to the pool present in the basin to create a more productive system (Kadlec 1979). Nutrient enrichment also takes place by windblown or water-moved soil, as has been shown in prairie wetlands (Adomaitis et al. 1967). The plant cover on the surrounding uplands therefore influences the rate at which such soil moves into the wetland, and thereby the rate of nutrient buildup. Increased nutrients also may enter the system in the form of plant detritus from upland or marginal areas.

Limited studies of primary production and nutrient cycling in the marsh system have provided a basis for some general concepts of the steps in the nutrient flow through a marsh system. Because of the typical anaerobic conditions at the bottom of a marsh, decomposition and mineralization of organic materials are slower than in terrestrial systems where the presence of oxygen speeds the process. The general scheme seems to be that the emergent aquatic plants that characterize marshes are the ones that tap the above-water-level oxygen and the nutrient pool stored in the soil substrate or water to build major organic structures (van der Valk et al. 1979). Submergent plants, which also may be rooted and obtain nutrients from the substrate, constitute a major proportion of the nutrient flow in shallow open marshes or lakes (Carignan and Kalff 1980). Such plants—emergent and submergent—are referred to as nutrient pumps because they move phosphorus and nitrogen from the substrate to the plant. Submergents and floating plants also may take nutrients directly from the water.

The major biomass of biological organisms in the marsh ecosystem usually is represented by the larger emergent plants. In marshes, this primary productivity (i.e., net biomass produced per unit time and area) is enormous, exceeding that of grasslands and equaling or exceeding that of tropical forests (Etherington 1983). For example, the metric tons of emergent plant per hectare per growth season in one study was estimated as follows: sedge = 10, reed = 21, and cattail = 27. (The U.S. tons of emergent plants per 2.5 acres per growth season would be estimated as: sedge = 11, reed = 22, and cattail = 30.) Such emergents constitute the dominant plant structure as well as the greatest production unit in the marsh.

All wetland plants eventually return nutrients to the water or sediments; the time that takes depends on the durability of the plant. Nutrients may be extracted by leaching and physical actions of water, by breakdown of plants due to herbivores, and by processing of debris by detritivores. Finally, decomposition results from use of the fine particles by bacteria and fungi. Ultimately these nutrients return to the category of available nutrients (de la Cruz 1979). Key nutrients for plant growth, nitrogen and phosphorus, may be tied up in the sediments or the detritus, which seems to limit plant growth and hence production of the entire system.

Experimental studies of nutrients in marshes have shown that wetlands modify many chemical parameters of water as it flows through the system (Kadlec and Kadlec 1979). Heavy metals are immobilized and often deposited. Many nutrients are trapped in the biomass and sediments, and thus we refer to the system as a nutrient sink. Most subsequently cycle through the components of the system, but some are lost through outflow.

Excessive enrichment occurs especially when phosphorus is added to a wetland through upslope runoff, and eutrophication results (Shindler 1974). Algae and floating plants that take nutrients directly from the water tend to dominate and may shade out other plants as well as reduce available oxygen for animals. Microscopic and filamentous algae, floating duckweeds, and watermeal dominate such systems.

CHAPTER 5
Habitat and Behavior Patterns among Marsh Wildlife

Although we will later review some of the major animals of marshes by taxonomic categories (e.g., insects, birds, mammals), it is important to recognize that the individual species of marsh communities often have evolved together and may form a distinctive and functional entity. Some of the characteristics of the components of the ecosystem were mentioned earlier, but before examining the marsh fauna, we should consider several additional ideas. In most cases, I will use examples of birds, because these are the ones that I know best.

Animal Adaptations

It is generally assumed that early life had its origin in water and that many species and groups are still restricted to that environment. Fish and many invertebrates such as clams and crustaceans live underwater, obtaining their oxygen and food there. But organisms adapted to shallow water had to evolve adaptive mechanisms for survival in times of drought, or the drying of a pond might totally eliminate a population or even a species. Thus, drought-resistant eggs, which respond quickly to reflooding, and short life cycles are characteristic of many species of microscopic protozoans, crustaceans, and insects (especially mosquitoes). Those insects such as dragonflies that have immature stages lasting several years are characteristic of more permanent water (Kendeigh 1961).

But clearly many of the abundant and conspicuous vertebrate forms such as birds and mammals did not evolve in water but are readapted to the aquatic system, probably because of its rich resources. Obviously, habitat adaptation requires major anatomical modifications for surface locomotion (swimming, wading) or for specialized flight within a restricted environment (hovering and diving by terns, vertical takeoff by ducks using small wetlands). And even perching in a marsh presents special difficulties. Imagine the typical sparrow trying to hang onto the slippery, vertical stalks of marsh reeds. Wrens and yellow-headed blackbirds do it neatly, using a "spread-eagle" technique with one foot on each of two stalks. For such reasons, fewer semiaquatic vertebrates than invertebrates are well adapted, and some have adaptive strategies that involve only using wetlands when convenient. Many simply must avoid the rich resources of this habitat because their specializations are for terrestrial habitats, such as grassland, shrubs, and trees.

Competition and Resource Segregation

A long-held concept of ecology has been that no two species may occupy the same niche and exploit the same resources (Odum 1971). But much ecological research today is directed toward the ways in which species do seem to accomplish this. Competition for resources brings about specializations that reduce direct conflict. The presence of competitors narrows habitat selection, and the variety of species in a community seems to depend on their ability to survive without excessive competition. Many diverse mechanisms have evolved: variations in seasonal use of foods or areas, different foods, similar foods in different habitats, various sizes of foods, feeding at different times of day or night, and so on. The result is that a marsh can have a large number of bird species, mostly secondary or tertiary consumers, that seem to make use of extremely abundant resources without severe competition, and some of which feed on different resources in different places in sometimes different ways, so that competition does not limit their presence. The waterfowl (ducks, geese, swans) are a good example. Swans and wigeon feed on submergent plants in water or on fine grasses along seasonally flooded shorelines. Geese feed on drier sites on sedges and grasses, and some feed on the tubers (snow geese) whereas others feed on stems and leaves (Canada geese). Many dabbling ducks are omnivores, consuming seeds and foliage usually in fall and winter (Bossen-maier and Marshall 1958) and feeding on more invertebrates in the pre-breeding period (especially the females, which need special nutrients for egg production) (Krapu 1974, Krapu and Reineke 1992). The apparent use of similar resources by dabbling ducks such as mallards and blue-winged teal in similar feeding sites has not been adequately explained, but often the sites have superabundant food resources. Inland diving ducks feed on bottom organisms such as snails, clams, and midge larvae in water depths that require diving. Sea ducks use different habitats and are almost purely animal feeders, taking some very large, hard objects such as mollusks (eiders), or fast-swimming but large prey such as fish (mergansers). Thus, the various groups are well segregated by their habitat selection as well as by their food, and many seem to feed side by side with little of the aggres-siveness seen among potential competitors. We shall examine some other examples of resource segregation as we later review some of the dominant groups of animals characteristic of a midwestern marsh.

Social Relationships

Interspecific social relationships are strongly influenced by the increased concentration of a species' population in the marsh. Some birds are dense colony nesters; some nest in a loose colony; others remain solitary, but may cluster because of the patchiness of suitable habitat. Egrets and ibises, eared grebes, and yellow-headed blackbirds are typical colony nesters. Nests of western grebes and some ducks may be found in clusters in suitable habitats.

Patterns of breeding behavior are closely tied to the clustering of nest sites, and although monogamy is still the common practice of marsh birds, the predictably abundant food resources permit multiple mating systems (polygamy) (Orians 1961). The males of several species of marsh blackbirds and of both marsh and sedge wrens seem to mate with as many females as they can accommodate in their territory. Possibly because the females can find an adequate amount of food for feeding the young, males can devote their efforts to territorial defense. But even within a productive marsh, nest sites do vary in quality (although we have difficulty quantifying what attracts a nesting female to a particular site), and some male redwings will establish territories in places where few or no females will nest. Other males will end up with four or five females.

The social system of a particular species may benefit several different species in the marsh community. For example, blackbirds, terns, and gulls are constantly alert to potential predators such as hawks, mink, or people; each has a special alarm call, and most engage in mobbing behavior. Such behavior not only alerts individuals of the same species, but seems to alarm other species as well (as any hunter or birder can attest!). An incubating duck on the nest becomes alert as a tern calls overhead, and hens with ducklings are even more wary. These interspecific warning signals clearly function for the good of all in the community.

Sounds of the Marsh

Birds and amphibians as well as insects create the sounds that we know as a marsh: One can be in a plowed field or a dense forest and know a marsh is just over the hill. Marsh sounds are not always pleasing, but harsh bird calls carry a long way over the noise of water and wind in reeds. Dominant vocalizations at night are the choruses of the leopard frogs and American toads. The mating calls of males attract females, and the marsh may be simply swarming with toads or frogs at certain seasons and in some years.

Different insects may be active at different times of night or day, and most produce buzzing sounds, such as those of mosquitoes or midges. Hordes of midges make an impressive drone, and those who believe them to be biting insects are understandably concerned. Fortunately, they are merely a nuisance during peak swarms, when they may coat everything, including human eyes, ears, nose, and mouth.

But it is the bird sounds for which a marsh is best known. In the Midwest, these include the raucous "poppycock" call of the male yellow-headed blackbird, the "creee-e-e" of the red-winged blackbird, the resounding "thunderpumping" of the American bittern, the winnowing series of chirps or chatter of rails, the sharp alarm calls of terns overhead, the "churking" of coots, and the various hornlike or laughterlike calls of several species of grebes.

Marshes bring diversity in many ways, and their sounds are among the most fascinating; though sometimes harsh, they reflect the richness of the marsh and demonstrate strategies for living together—yet apart.

The Marsh Edge

The transition zone between two plant communities may form a broad ecotone, as in the case of the prairie and the deciduous forest in the Midwest, or it may be abrupt, forming a discrete edge. Either type of edge strongly influences animal distribution, as noted among birds, but the interface between marsh edge and upland is notable for other reasons. First, it is a dynamic front, where many species of plants characteristic of moist soil or periodically flooded zones are found. The plants may be shorter, and richer in species, than those growing in either the marsh itself or the upland zones. Animals too may be diverse in this zone, depending on the mix of plants and the water level. Quite open mudflat areas may be frequented by killdeer or other shorebirds and waders. Frogs of several species use wet meadows adjacent to marshes. Blackbirds do likewise, and bitterns and herons may catch meadow mice as well as insects in such areas. Some bird species such as swamp sparrows and yellowthroats are in their prime habitat here; this is where they nest, although they may stray some distance into the uplands or into the marsh to forage. Some terrestrial blackbirds fly hundreds of yards from the marsh to seek insect foods for their nestlings, although adult males and nonbreeders may feed on waste agricultural grains as well. Some species such as the mallard and blue-winged teal nest in the uplands or at the marsh edge but feed in the marsh during the breeding period. They are plagued by such terrestrial predators as striped skunks, raccoons, and ground squirrels, but their persistent nesting behavior helps to compensate for high losses.

Mink and raccoons concentrate along the marsh edge, where they can move in the not-so-dense cover and capture crayfish, mice, and frogs. Deer have well-defined trails leading to the marsh, where they drink, and they find shelter in the willow thickets or in the reed or cattail beds, but they prefer the drier sites for bedding down. Thus, a strong evolutionary influence of the upland-marsh interface exists because marshes often go dry: Those adapted to the edge are always present, whereas aquatic specialists must leave, or die.

But the most vital aspect of this interface may be biochemical, for here the marsh gains additional nutrients from rainfall, drainage, and seepage through upland soils. Nutrients from decaying plants and animals eventually find their way downward to the marsh basin. Masses of swallows that feed over both land and marsh roost in the marsh vegetation at night, "whitewashing" large areas and adding processed nutrients to the ecosystem.

CHAPTER 6
Dominant Animals

A brief review of the dominant animals, concentrating on the more conspicuous wildlife of midwestern marshes, will document the impressive variety found in even a small area. The animals' use of this unique system dramatizes habitat selection and segregation among closely related species, showing the roles these forms play in the functioning of the ecosystem. We will consider the fauna by taxonomic groups, but without neglecting habitat selection, niches, relationships, and species associations. We will start with the most conspicuous, the birds, and end with the less conspicuous but perhaps most important in the system, the invertebrates.

Birds

Although generalizations are inherently dangerous, some general patterns may help the reader to place species in perspective with others and to appreciate the evolutionary forces that have induced habitat use, breeding behavior, and use of space by birds. The marsh has been a strong attraction to bird groups not only because of its food richness but also because it provides nesting, resting, and feeding sites protected from ground predators. Various bird groups have adapted anatomically to such aquatic situations to the extreme that they are less efficient on land (loons, grebes, inland diving ducks); others use upland as well as aquatic habitats and have added flexibility, but certain disadvantages too (dabbling ducks, some waders, rails); and numerous species use the marsh edge as the focal point of their feeding and breeding (swamp sparrows, sedge wrens, red-winged blackbirds).

Actually, some whole orders (the major taxonomic divisions within the class of birds) have specialized in aquatic habitats, and various subgroups (families, genera, species) have adapted in varying degrees. We will discuss variation in closely related species (Fig. 3), considering generally how they use different habitats and foods (Weller and Spatcher 1965). Most breeding birds are carnivores or omnivores, with only a few large herbivores in the marsh. As we shall see, use of marshes, habitats, and foods varies seasonally as influenced by needs for reproduction or by availability in the system. Well-adapted aquatic species are limited to marshes or lakes year-round even if they migrate long distances, and they use few if any terrestrial foods (some inland diving ducks, grebes). Marsh-edge species also tend to seek out similar habitats all year, but they have greater flexibility, seeking food in uplands as well as in marsh areas (geese, blackbirds).

It is impossible to consider all the birds in a typical marsh, but a few examples of dominant groups will demonstrate the pattern. Loons (or divers) normally use the deeper, more sterile water of large lakes, but common loons may use marshes in northern Minnesota and Wisconsin where the water is deep enough to hold a fish population. Grebes of several species prefer marshy areas, especially during the nesting season, but all may feed along the seashore in winter. Grebes are so specialized with their rear-mounted legs that they walk poorly on land. The largest North

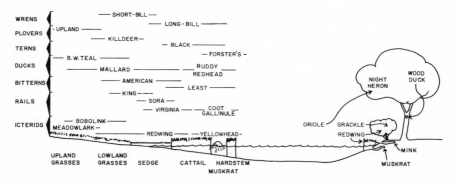

Figure 3. *Competition for nest sites and associated food commonly brings about the segregation of closely related species into adjacent habitats with reduced overlap. Whole families of birds may be tied to wetlands, such as herons and bitterns, or rails, whereas other families of birds (Icterids) have numerous upland representatives as well as those adapted to marsh (Weller and Spatcher 1965). Thus, in a specific vegetation type or structure (e.g., open-water edge), or in a marsh that is entirely of that type (e.g., deep marsh), one can predict the species to be found there by their habitat preference.*

American species, the western grebe, uses open areas and nests in extensive beds of reeds and other emergents. Red-necked grebes, horned grebes, and eared grebes use different types of wetlands, according to preferences for pond size and density of cover (Faaborg 1976). Eared grebes are especially colonial but are restricted in distribution by their need for suitable habitat. The solitary pied-billed grebe is the most widespread, tolerating dense to quite open situations, and is very responsive to new habitats, perhaps because it is versatile in food selection, using invertebrates, fish, and amphibians taken mostly underwater (Provost 1947). Although all grebes build low, wet nests made of submergent or emergent plants, the pied-bill is best known for its nest, usually a floating mass of rotting vegetation that is inconspicuous; its egg success is quite high (Sealy 1978).

Ducks range from upland nesters that use water only for feeding, loafing, and brood rearing (mallard, northern pintail) (Duebbert and Lokemoen 1976) to diving forms like ruddy ducks that rarely leave the water. Ruddy ducks walk poorly on land, nest over water, and obtain foods almost entirely by diving. Most ducks are omnivores, but during laying, the females use a pure invertebrate diet essential for egg production.

The rotting cattail stems of which this pied-bill grebe nest was built probably warm as well as humidify the eggs.

One duck that specializes on animal foods year-round is the northern shoveler. With its spatulate bill and fine lateral lamellae (strainers), the shoveler is the "whale" of the duck world, straining out fine plankton that less-specialized bills could not efficiently utilize. It is not so specialized that it cannot consume seeds and larger invertebrates like snails, however. One duck, the American wigeon, specializes on leafy plant foods, and feeds on various succulent submergent and emergent plants. Its bill is shorter than that of most dabbling ducks and more gooselike, and it also grazes in the uplands during winter. The canvasback has a long, sloping bill well designed for digging out the buried tubers of sago pondweed that it must reach by diving.

But the geese and swans are the major wetland herbivores, and it is not surprising that they are the larger-bodied birds, with long necks for reaching remote foods (they are the "cattle" of the waterfowl world). No specialized divers occur among geese and swans (although fulvous whistling ducks, which are closely related to the swans, dive well), but considerable variation exists in habitat choice, from dry-land geese that take their young to pond edges to feed and into the water for safety, to swans that nest over water and feed mostly on submerged aquatics.

Of the heronlike birds, the bitterns are perhaps the most closely associated with water. The most aquatic bittern is the least bittern, which builds a delicate, solitary nest several feet above water in cattail or bulrush, and feeds on frogs and other marsh animals of suitable size (Weller 1961). It is rarely observed from shore, and, unless one knows its dovelike call, it can be present in abundance and never recorded by "birders" who choose not to get their feet wet. The larger American bittern is found mostly along the marsh edge, where it feeds on mammals and snakes as well as frogs, but it may penetrate the cattail or reed zone to build a solitary nest over water. However, upland nesting by bitterns is common in some drier prairie areas.

Herons, egrets, and ibises generally nest colonially in wetlands, also feeding there or in nearby wetlands. Herons and most egrets feed on fish, frogs, and invertebrates in shallow marshy areas, and the black-crowned night heron occasionally even swims to catch fish. But not all waders that nest in the marsh feed there; ibises and storks nest in water-protected sites such as shrubs or emergents of deep marshes but feed in the wet meadows and uplands on animals. The pioneering cattle egret reached the United States from South America in the early 1950s, and feeds in the uplands where it catches arthropods disturbed by grazers. Depending on the time and place, great blue herons may nest in emergent vegetation or in trees in upland settings, but the best protected nests I have seen were in sprawling beds of prickly pear cactus on dredge-spoil islands along the intracoastal waterway in Texas. Clearly, marsh birds are flexible, and localized behavior patterns are common.

Terns are the insect gleaners and shallow fisherfolk of the marsh system (Bergman et al. 1970). Black terns feed mostly on insects at the water's surface, whereas Forster's terns feed on small fish in larger pools. Black terns nest on low, soggy debris of old muskrat lodges, or build similar low nests of wet (old or new) vegetation (Cuthbert 1954). Forster's and common terns may build larger and drier nests than do black terns, but they most often nest on high, dry muskrat lodges (Nickell 1966). Forster's terns seem to get along with muskrats fairly well, as their nests are rarely lost to muskrat activity, possibly because muskrats do little lodge building at this season of the year (Weller and Spatcher 1965). Both tern species may have grouped nests when conditions are good, but rarely are they found in huge colonies, as is common among other gulls (Anderson 1965) and terns.

The rail family has representatives that span all the habitats of a typical marsh. The most aquatic representatives are the American coot and the common moorhen, which nest over water and feed their young on aquatic invertebrates. Coots are excellent divers, and older young and adults feed on submergent vegetation, whereas gallinules are more surface pickers and tend to occur in or close to heavily vegetated areas. Coots will graze on upland grasses in winter and during migration. Virginia rails and soras often are present in large marshes and often are heard, but not seen, because of their preference for dense vegetation. Virginia rails tend to use wetter sites in more robust vegetation, often nesting in shallow stands of cattail. Soras favor sedges in shallow water, but they also can be found well out in wet areas. The largest rail, the king rail, generally nests and feeds in marsh edges or upland areas and has, therefore, suffered more severe losses of habitat.

Blackbirds are conspicuous and abundant residents of marshes, with some species essentially restricted in nesting to flooded marsh emergents. In the Midwest, the yellow-headed blackbird is the central-marsh resident, and the red-winged blackbird reaches its peak density along the edge (Weller and Spatcher 1965). During the breeding season, young are fed almost purely on invertebrates, but adults regularly feed on grain and other seeds as well as insects (Orians and Horn 1969, Voigts 1973). Redwings are a highly social species, and often are polygynous: Males in especially favored habitats may have two to five nesting females, successively if not simultaneously.

Two wrens are common in, and nearly restricted to, marshes over a wide geographic area. Again, they overlap little. The marsh wren nests over water in robust vegetation such as cattail or hard-stemmed bulrush, whereas the sedge wren favors tall wet-meadow grasses and even wet-prairie vegetation. Both are insect feeders, gleaning the vegetation more than feeding in water. Polygyny also is common, with large numbers of dummy nests being built by the agile, vocal, and hyperactive male to attract use by multiple female mates.

Several other passerines favor marsh edges, such as the swamp sparrow and yellowthroat. Swamp sparrows forage in the marsh edge but may use taller vegetation in damp sites for song perches and nest sites. Yellow-throats may nest in much drier upland sites as well but seem to prefer wet areas.

Obviously, birds are an adaptable group. Any plant life form has its characteristic cluster of bird species, with little overlap of closely related species—and seemingly little competition for food, cover, or other resources (see Fig. 3). However, if for some reason one species is missing in an area, such as occurs in the timing of migration of red-winged and yellow-headed blackbirds, the first to arrive (redwings) take over much of the marsh but later may be displaced by the more dominant yellowheads. Hence, there is still some latitude in habitat use by many species, and more research is needed to fully measure and assess these complex interactions.

Most of these breeding marsh birds feed their young animal foods, and probably have adapted to the marsh because of this rich resource that includes invertebrates, amphibians and fish. But in late summer and fall, when seeds are most abundant, many young birds switch to seeds or leafy vegetation and exploit the seasons' greatest production. Birds en route to and on wintering areas also take mostly vegetable material in their diets.

Mammals

Relatively few mammals are truly marsh specialists. This is a seemingly major adaptation and one that cannot be complete; the effects could be disastrous for a nonmigratory population because of the seasonal and annual fluctuations in water levels and marsh quality. Beavers compensate for fluctuating water levels by damming streams and creating their own stable pond. Otters adapt through maintenance of strong terrestrial behavior and through their flexibility in using streams, lakes, or ponds; but marshes must have large fish or amphibians to attract them. However, otters are very abundant in coastal fresh and salt marshes of the Gulf Coast, where they feed on crayfish, crabs, and mammals as well as fish (Chabreck et al. 1982).

The presence of fewer species of mammals and of fewer mammals overall is more than compensated for by the relative importance of these specialists in the functioning of the system. The widespread muskrat and the nutria (introduced from South America) constitute the major herbivorous wetland specialists. Interestingly, these animals probably evolved in two of the most extensive marsh-dominated areas of the world, the Prairie Pothole Region of the United States and the pampas marshes of Argentina. But both herbivores are fairly adaptable and are found in a variety of aquatic and wetland habitats. All use a great variety of vegetation as food, but all grazers favor either growing shoots because of their concentrated nutrients or rootstocks and tubers that store nutrients for subsequent growing seasons. Muskrats and their close relatives, the round-tailed muskrat (or water rat) of Florida, and the nutria build ball nests for rearing their young; these balls are made of shredded vegetation and are lodged in sturdy vegetation above the water level. Beavers also get into large and fairly stable marshes (Beard 1953), either because of regional preference or as stream populations overflow. In such situations, beavers may build lodges of and eat cattail instead of willows or cottonwood, but I know of no studies of their survival or reproductive success in such situations.

Muskrats and beavers influence marshes more by their cutting of vegetation for lodges, storage, and nests than by their cutting for food consumption. In the North, these lodges provide protection against the cold and heavy snow, as well as a food supply that may be exploited in late winter and early spring, when resources are at their annual low. Muskrat lodges are built quickly in the late summer and early fall when the vegetation has reached its peak. Some are six feet (1.5 meters) high and fifteen feet (4.6 meters) in diameter, and terrestrial plants may grow on them. They may be abandoned after one season and settle into the water from autumn rain and winter snow. Enriched by bird and muskrat droppings, they may become germination sites of semiaquatic plants. The pools formed where lodges were built remain open for a long time in some situations, perhaps because of their high organic content and lack of aeration when flooded. These open pools are favored places for a variety of birds.

More species of carnivores use marshes, but they are less specialized than are the herbivores. The otter is more truly aquatic, frequenting deeper

The muskrat lodge, a product of the marsh and of a marsh specialist, the muskrat, is in itself a fascinating "island" utilized by many forms of life. In addition to terns and ducks that may nest upon it and muskrats, mice, raccoons, or mink that may lodge within it, there are mites and insects that live in the plant stalks, eating, decomposing, preying on, being preyed on, and so on. "Scuds" and snails below the water level devour rotting vegetation, and minnows and bullheads find food and shelter in the passageways.

water, where it utilizes fish as a major food. The mink is a more typical marsh associate, often tied to muskrat populations, although mink inhabit streams and lakes as well. But especially during the breeding season, the easiest place to find mink is along the edge of a marsh. Trails along the shoreline of wet meadow are conspicuous, and remains of prey and telltale droppings denote catches and feeding sites. In an old muskrat burrow at the base of a near-flooded willow tree, or in a muskrat lodge itself, piles of droppings, feathers, wings, and mammal fur tell of a litter of young. Muskrats are perhaps the major food of mink in these situations, but crayfish, frogs, fish, birds, and small mammals also are taken.

Several other mustelids frequent marshes, with short-tailed weasels being reasonably common in the northern states. Least weasels, the smallest of the group, have one of the most northerly of ranges and even frequent Alaskan tundra wetlands.

Although raccoons often are found in terrestrial and even woodland habitat, some populations, at least in summer, use marsh edges and muskrat lodges. More than once I have encountered a raccoon curled up in a muskrat lodge far from shore. Trails along the marsh edge are usually littered with the remains of crayfish—a favorite of raccoons—suggesting that these remains are mainly prey items captured by raccoons.

Meadow mice and short-tailed shrews frequent marsh edges, and our small-mammal trapping in northwest Iowa demonstrated higher densities in wet meadows along marshes than in higher and drier bluegrass and prairie grasses. Meadow mice do swim and occasionally even nest in the lodges of muskrats or in the bases of duck nests. Whether this behavior (and the resulting population density) is forced by water fluctuation or is a localized habitat selection trait is uncertain, but I know of no real measure of population densities of any marsh mammals other than muskrats and nutria.

This variety of mammals exemplifies the well-known food pyramid common to many natural communities. Plant material is the most abundant and easily obtained food source, so herbivores such as muskrats and meadow mice are the most numerous species present. Raccoons are omnivores that selectively use both plant and animal foods, and food availability is reflected in their commonness. Shrews, weasels, and mink are strictly carnivores, feeding mainly on herbivores or omnivores, and thus are less common because of the lesser abundance of food sources and the energy they must expend to pursue and capture prey.

Fish

Relatively little is known about fish in marshes, but the marsh environment limits the kinds that can live and, especially, reproduce there. Because of the shallower depths, reduced wind action, higher water temperature, reduced oxygen, and sometimes more turbid water, larger fish are those that we associate with such highly eutrophic systems: bullheads and introduced common carp. But water depths vary seasonally, and oxygen needs and tolerance vary with ages of the fish. The northern marsh canoeist is impressed in the spring when a streak underwater denotes the presence of northern pike that enter lakeshore marshes to breed. Later in the season, predation on ducklings is another measure of their presence (Lagler 1956). But young pike move out of the marsh and spend most of their adult lives in more open water. For northern pike, this relationship of marsh to lake is so vital that fish populations decline when water levels prevent spawning and rearing in such areas (Forney 1968, Kleinert 1970). Other fish also come to spawn, and young (especially the preyed upon) find protective cover in the emergent and submergent vegetation and may remain in these rearing areas until quite mature (Heding 1964, Priegel 1970).

But the larger predatory fish probably are not as important in the energy flow in a marsh as they are in lakes, and the predatory niche often is filled by herons. Although factual data on fish populations are difficult to obtain in a marshy situation, it has been done with various drop traps or scoop nets (Kushlan 1974). Clearly much more work needs to be done with this group of vertebrates. Masses of brook sticklebacks and fathead minnows presumably are the major underwater omnivore and animal consumers in the marsh system (Peterka 1989). Certainly when larger predatory fish enter the marshes, even if briefly, they feed on these smaller fish readily. The importance of fish of all sizes to storks and herons is especially well understood in the Florida Everglades where water cycles create drying pools that make fish more available to such predators; these conditions dictate if and when these birds will nest (Browder 1978).

Amphibians

Salamanders, frogs, and toads represent specializations from the most to the least aquatic amphibians that use marshes. But as with fish, their niches and habitats differ with age. Larval amphibians are aquatic and feed on minute plants and animals, but as adults, they gradually shift to animal food of larger sizes, including minnows and invertebrates. The adults may be totally predaceous, but their range in foods probably changes by habitat as well as by season.

Salamanders are highly aquatic; however, some species such as tiger salamanders move overland in large numbers at times, but probably more because of breeding activity or wintering site selection than because of food. These animals can be major food items for birds such as white pelicans that use marshes and shallow lakes to gather food for their young (Lingle and Sloan 1980).

Frogs, such as the common leopard frog, use both shallow aquatic areas and wet meadows, where we have often taken them in traps designed for small mammals. The giant bullfrog of warmer southern and eastern waters is found in deeper water and is a significant predator on larger prey items, taking other frogs and even ducklings. Toads, of course, are more terrestrial, turning up in very dry sites long distances from water, but in the spring, it is the marsh where they chorus, mate, and breed. The eggs hatch, and the young grow in the warm, food-rich, and protected shallow waters. As young, schools of tadpoles are conspicuous along marsh margins, seemingly so vulnerable that one wonders how they reach maturity. But the jab of a stick, like the jab of a heron, produces a coordinated predator-escape response that demonstrates how schooling benefits the majority. Amphibians are prime food sources for all the larger predators such as mink, raccoons, herons, bitterns, and fish.

Reptiles

Three major groups of reptiles occur in wetlands in various parts of the country. Of these, turtles are the most widespread geographically and the most common in terms of both number of species and size of total population. Species vary from place to place, but the sight of painted turtles basking in the sun is as common in marshes as it is in lakes—only the sunning site is more likely a muskrat lodge than a rock or log. Although painted and mud turtles are common, less often seen is the massive, predatory snapping turtle that slips slowly and silently underneath its prey and helps to reduce overpopulation in a local area. Its prey includes fish, frogs, and birds.

Snakes are also fairly common in marshes, but perhaps less so than in swamps and river wetlands. In the northern marshes, the only regular resident is the garter snake, which seems mainly to patrol the marsh edge where it takes eggs and young from red-winged blackbird nests, but it does not fear water and occasionally is found basking on a muskrat lodge. In more southerly areas, numerous species of water snakes dominate, but there also are poisonous water moccasins. Water snakes feed on fish, frogs, and other vertebrates opportunistically.

Southern marshes and other wetland types also are known for the largest and most feared of all reptiles, the alligator. Each has its "hole" that seems to be dug out and retains water even in the drier periods. Its role as a major predator on birds and mammals must be significant, yet herons still wade, coots still swim, and muskrats still nest adjacent to alligator holes.

Invertebrates

Because these small to microscopic animals are so diverse, representing thousands of species, I mention here only a few of the dominant, macroscopic groups that play significant roles in the marsh system (Kendeigh 1961, Needham and Needham 1941, Pennak 1978). Several uncommon but unique groups occur such as the freshwater jellyfish and freshwater sponges, but it is the ubiquitous forms, regardless of size, that are principal food resources for the wide range of vertebrate predators already discussed. They are the core of the food chains and webs in the marsh. Many are predators themselves, such as the dragonfly nymph that catches minnows or the adult dragonfly that catches mosquitoes on the wing. Tiny swimming protozoans and crustaceans are trapped by bladderwort plants; large crustaceans also catch all kinds of small crustaceans and insects; and a wide range of predaceous insects are specialized for catching food either on the surface or under water.

The most conspicuous aspect of invertebrate life associated with marshes is the myriad of insects that we consider both a nuisance and, at times, a serious problem. Without a doubt, the most abundant group are the true flies, the Diptera, which include midges, mosquitoes, and crane flies. The aquatic larvae of mosquitoes occur on the surface of shallow edges and are eaten by fish, frogs, and other invertebrates. Because mosquitoes do breed in the shallow waters at the edge of marshes, and most people experience only this edge, marshes have an undeserved reputation for being unbearable because of the abundance of mosquitoes. Depending on the area and the species involved, wet meadows, oxbow swamps, and wet forests are far more important mosquito-production sites. This misinterpretation is perpetuated by misidentification of marsh flies (midges) as mosquitoes. This large family of insects is one of the most important in marshes all over the world. The adults swarm in unbelievable numbers and can be photographed in giant swarms against the setting sun. As is true of many marsh organisms, it is perhaps the generally unseen larvae of the midges that do most to support the system. Because of their rich red color, these are called "bloodworms"; they are found submerged in bottom soils and organic debris, serving as food for fish, frogs, and diving birds. When pupae surface and emerge as adults, they are exploited as well by surface-feeding birds and fish. Swallows, ducks, terns, and the smaller and more agile gulls eat continuously without seeming to dent the numbers.

Mayflies occur in deeper marshes and leave their abundant cases floating on the surface as they fly away; their swimming larvae are abundant and rich food resources. The larger dragonflies and damselflies are characteristic sights of the marsh—the adults darting here and there—stopping so close, yet too fast to touch or catch. The larger dragonfly nymphs go through many growth stages (naiads) and require one to four years for maturation. Hence, they are expected only in the more permanent waters.

Swarms of midges hover over the beach ridge of the Delta Marsh at sunset, Delta, Manitoba.

The true bugs (Hemiptera) are represented by forms as diverse as water striders that flit about on the surface tension of the water and predaceous backswimmers with prominent eyes and enlarged, oarlike legs. They may occur on the marsh bottom and come to the surface "belly up" for air, which is then carried with them externally. The most impressive "bugs" are the giant water bugs that may be nearly three inches long; these bugs prey on tadpoles and small fish. Also well known are the water boatmen that, though they swim with darting movements like predators, tend to eat fine algae and bottom debris.

Early summer on the marsh can produce hordes of mothlike caddis flies. Many of their larvae build cases of bits of vegetation, other organic materials, or sand. The larvae are a choice food item for some ducks that seem to be able to extract the larvae without eating the cases. A number of aquatic beetles occur, the best known being the predaceous diving beetles, some of which are nearly two inches long. But many species exist, and their larvae are abundant on the bottom or in submerged vegetation. Better-known beetles are the "whirligigs" that occur in colonies at the surface, moving together in sweeping circles and scattering at the slightest disturbance.

Among the most prominent of invertebrates are the crustaceans, which can be encountered even in food-poor marshes. Many of these are tiny drifting or free-swimming forms termed plankton; these include the water fleas such as daphnia and the copepods such as cyclops that challenge the unaided eye for identification. All are vital to the efficiency of food chains, for these are the animals that tap the superabundant algae and protozoans of still smaller sizes. The larger, more mobile fairy shrimp can dominate wetlands of a more temporary nature during cooler times of year, and constitute important food resources for fish, amphibians, and birds with efficient straining bills. Like insects, many crustaceans are associated with the larger submerged aquatic plants that provide food, substrate, and protection from predators (Krecker 1939, Rosine 1955). Examples of these crustaceans are the scuds or sideswimmers and some species of water fleas (Quade 1969). The minute seed shrimp or ostracods may use such vegetation but also are found in decaying vegetation and bottom mud (Pennak 1978). Other bottom-dwelling crustaceans are the aquatic sow bugs and several species of crayfish. The latter are large crustaceans that exploit various habitats and ingest both plant and animal foods (Momot and Gowing 1978). Most are bottom forms, or *benthos,* and the shoreward species build conspicuous mud tubes from their underwater burrows. Shoreline crayfish suffer heavy predation from mink and raccoons; the more aquatic forms are used by predaceous fish.

Snails and other mollusks can be extremely abundant organisms in most marshes, and are mainstays in the diet of many vertebrates. Fingernail clams can be superabundant in silty areas, but seem more common in open, river pools and lakes than in marshes. One study on a Mississippi River pool reported 40,000 fingernail clams per 10.8 square feet (one square meter) (Gale 1975)!

In addition to their use as food (Schroeder 1973) and their conversion of plant materials to animal protein, the most essential role that invertebrates play in the system is as shredders and detritivores; they break down plant material to a size and structure where bacteria and aquatic fungi can further process these products. In the absence of these detritivores, the one-way production line would stop when all nutrients were tied up in organic material that could not be recycled. This happens in cold, acid, or polluted areas when decomposition does not keep pace with production potential—and sterile conditions result.

CHAPTER 7
Habitat Dynamics

Seasonality and Wildlife Responses to It

Perhaps the most dramatic form of habitat change, seasonality, is so common that we take it for granted. Seasonal change is especially conspicuous at high latitudes (both far north and far south), with cold being the dominant physical force. But where seasonality of precipitation is great, it can be just as important in habitat change as is cold. Often, the two are interrelated.

Our studies of tundra wetlands (only part of which can be termed emergent wetlands) documented the extreme case of seasonal variation affecting a habitat, where in the short summer season only about twenty-five species of birds breed in the wetlands and a few more occur in the uplands (Bergman et al. 1977). All birds in both habitats, except willow ptarmigan and snowy owl, move out during the bleak winter, and lemming availability determines whether owls stay or not. The same phenomenon occurs in the prairie pothole marshes of the Dakotas, Minnesota, and Iowa, except that a greater number of species is present in summer. In one North Dakota marsh, thirty-seven species nested during the summer (Krapu and Duebbert 1974). Pheasants, horned larks, and a small number of woodland birds remained in the uplands throughout the winter. Moreover, a still larger number of migrants use these marshes during their annual stopovers.

Obviously, a warm-climate marsh need not experience this kind of seasonal change, yet changes do occur, owing to annual cycles of water avail-

49

ability, plant growth, insect production, and use of the marsh by mobile birds. In one study of bird species using a southern shrub wetland (Louisiana), breeding birds numbered only seven species, but seventeen wintering species occurred (Ortego et al. 1976). During spring and fall, there were many more migrants. Breeding birds are tied to a marsh; hence, conditions there may restrict breeding. But flocks in the nonbreeding season are mobile and can utilize diverse foods where they are most readily found. Such adaptability not only is energy efficient, but maintains the maximal population of a species.

Like other birds, most marsh birds return in the spring to nest in last year's vegetation, but if this vegetation is missing for some reason, most birds probably move elsewhere. However, in several years of observation in the marshes of northwest Iowa, it became apparent that members of some species, yellow-headed blackbirds and American coots, remained in flocks in the marsh and nested later, when new plant growth appeared. Some of this behavior may be due to late migration by young birds, but habitat conditions also play a major role. Characteristically, yellow-headed blackbirds' nests are made of dead cattail leaves woven on old stalks, whereas these late nesters attached their nests to green shoots. Often one shoot grows faster than another, causing the nest to pitch over and the eggs or young to fall out. We suspect from color patterns that these birds are yearling females that nest late and probably are less successful. Delayed nesting also seems to occur in American coots that build nearly floating nests at water level; some may be awaiting growth of suitable nesting cover because late-season nests often are built of green foliage that is less buoyant than dried vegetation from the previous year. However, late nests built of green materials seem less buoyant in heavy winds, and the nest success probably is reduced.

Although birds and a few insects react to seasonality by mobility, relatively few marsh mammals migrate. Northern muskrats survive severe winters by lodge construction and food storage plus local movement to marshes of freeze-proof depth (Errington 1957, 1963). This behavior limits which marshes are suitable overwintering habitat for muskrats. A lodge is not impenetrable to a fox, but muskrats can escape underwater, making digging a probable waste of time for the predator. In times of drought, muskrats can suffer severe losses to such terrestrial predators (Errington and Scott 1945). Thus, seasonality affects what the muskrat can do and how it does it.

The effect of seasonality on truly aquatic forms is even more striking, because seasonal change causes the physical conditions to which the fauna respond in the evolution of their life cycles. Annual crops are the rule, however, for the majority of marsh animals of shallow marshes, because chances are good that marsh conditions will not be the same next year. Dragonflies, salamanders, and fish that require several years to mature are associated with more permanent marshes or ponds.

Succession and Other Changes

Our general concepts of natural changes in vegetation (and obviously associated animals) have been conditioned by long-term views and by our familiarity with terrestrial systems. Ecologists understandably have been most interested in piecing together past history, especially since the ice ages. There are, as a result, a few classical studies of plant and animal communities of ponds of various ages formed along old beach ridges, or of the various stages (and ages) of pond-bog succession as seen in northern coniferous forests (Kendeigh 1961). The texts tend to emphasize such successional trends, showing how a shallow lake eventually is filled by sediments and organic matter and becomes marshlike or boglike and eventually is grown over by the dominant local terrestrial vegetation (Fig. 4) (Buchsbaum and Buchsbaum 1957). The wetland aspect of this successional sequence is of importance here. Lakes that are clear and deep tend to be low in productivity (Reid 1961); they become sedimented, less clear, enriched with nutrients, and higher in basic productivity as part of the succession created by natural filling with rich soils and organic matter. The rate of such eutrophication depends on many factors of the surrounding uplands, as well as the depth of the wetland, the nutrient base stemming from its substrate, the rate of inflow and sedimentation, and the climatic regime (Livingston and Loucks 1979). These aspects of long-term succession also are factors in short-term changes, with which we are most concerned here.

Animal succession associated with long-term changes in plant life and water quality is reflected in the different stages that a body of water undergoes, from pond to grown-over marsh or bog. Thus, the deep and clear pond may contain fish that favor clear water, open-water birds such as loons that feed on fish, and invertebrates that require relative permanence. Later phases include the dense submergent stage with floating-leaf vegetation, marsh-loving birds, and invertebrates influenced by an abundance of detritus. Eventually, the moist terrestrial vegetation in the final stage

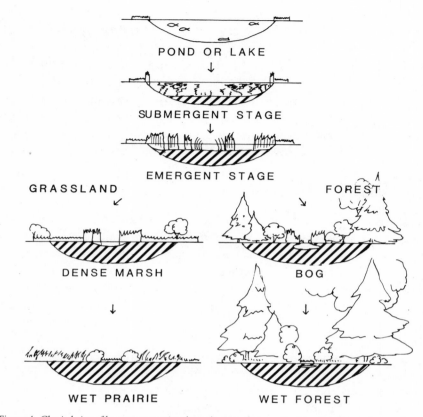

Figure 4. *Classical view of long-term succession shows the stages from pond or lake, through emergent marsh stage, to local climax terrestrial vegetation. Because of natural deepening processes, this may not occur at all or might involve thousands of years. See text for discussion of short-term patterns.*

demonstrates an animal fauna similar to that of the uplands (Buchsbaum and Buchsbaum 1957).

Although this general trend may be true over thousands of years, it may not always occur, and it may not be a simple one-directional successional pattern (McAndrews et al. 1967). There may be periodic reversals toward deeper water, or to drought or drainage that suddenly produces terrestrial conditions. The trends as well as the habitats are dynamic, forcing animal residents to be equally responsive, to move, or to die.

This classical concept is based on the theory of directional and almost predetermined long-term change in plant communities toward a climatically controlled climax vegetation (Clements 1916). Current concepts of marsh succession (van der Valk 1981) are based more on Gleason's concepts (Gleason 1917) of site variation, randomness, and plant life-history strategies with water as the major driving force. This viewpoint better explains the observations of short-term change that we see during drought-produced drawdowns in wetlands.

Because they have been less emphasized in textbooks, because they are more visible in a short-term time frame of three to ten years, and because they dramatically affect the wildlife of any discrete marsh unit, I want to detail some of the kinds of short-term vegetative change and wildlife-induced changes, as well as wildlife responses to these changes. My comments are based on observational studies, in which descriptions are made of one area annually for several years to document changes from which we infer successional patterns (Weller et al. 1958, Weller and Spatcher 1965), and experimental studies, in which manipulations of water level were used to recreate these stages and demonstrate causal factors (Weller and Fredrickson 1974). Some general concepts will set the stage for appreciating the dynamics of wildlife populations, the human influences on wetlands, and the necessity of management strategies.

Two forces dramatically influence change in marsh vegetation and associated animal communities. The most crucial is changing water depth, which, as indicated earlier, is the deciding influence on what plants grow where, and on what plant life forms (i.e., emergent or floating-leaf) dominate a particular marsh. Changing water depths also create stress for established plants, with deep water reducing the growth of even well-adapted emergents, and shallow water (and especially dryness) reducing growth of all but the wet-meadow or edge species. Thus, vegetative growth and plant distributions are dynamic, with rapid spread in some years and reduced growth or even dieback in others.

The second major influence on marsh vegetation is a product of the activity of herbivores such as muskrats (Errington et al. 1963), introduced nutria, or occasionally beavers that "eat out" large areas of vegetation used for lodges and food. Because of the physiological nature of most emergent plants, such cuttings will not regrow well if they are flooded by even an inch of water—and they will die in a year or so for lack of oxygen in spite of much stored food resources in the rhizomes and tubers (Nelson and Dietz 1966, Weller 1975a). If such cuttings are not inundated, or if the water level actually declines, regrowth is usual—though not always with the same density or robustness, since their photosynthetic ability is reduced and spring growth is based initially on nutrients stored in rootstocks.

Because water levels are usually a direct response to rainfall, and since rainfall patterns vary year to year and geographically, changes in marsh water levels are common in the Prairie Pothole Region (Kiel et al. 1972, Trauger and Stoudt 1978) and occasionally occur even in more stable climates in the northern and eastern forests. The interaction of these two forces thereby influences the vegetation and the wildlife of single marshes or even of marshes of major geographic regions. Whole areas may be dominated by wetlands in a certain condition; all may be shallow and meadowlike or flooded and semiopen. Arriving migrant birds that find conditions unsuitable in their natal marshes or prior nesting sites may shift to other wetlands in the same region or to different regions where conditions better meet their specific habitat needs of cover, food, or nest sites. To appreciate how variable these habitat conditions may be over a matter of a few years, let us first examine a common pattern in a midwestern marsh.

Short-Term Vegetational Changes. If an open marsh or shallow lake goes dry, the bottom sediments are exposed. Organic debris that has settled in the water for several to many years dries and decomposes through the action of invertebrates, fungi, and bacteria; invertebrates, turtles, and fish die and decompose; toxic substances stored (Cook and Powers 1958) under reduced oxygen conditions deteriorate. Soil structure is modified by alternate wetting and drying, and, finally, cracking that goes deep into the substrate layer. In some cases, soil may be blown from the basin, deepening it. In other cases, the wetland may form a catch basin for new and rich soil blown across the prairies or fields from unvegetated uplands (Adomaitis et al. 1967). The impact of such soil deposition on the wetland depends on the amount. Modest additions may act as fertilizer as well as build up substrate after shrinkage has resulted from decomposition of organic materials during dry periods, but some wetlands may be filled to a level that kills existing plants and induces a more shallow wetland plant community. Such wetlands would serve an important function of erosion control, but the habitat is changed dramatically for aquatic organisms.

Surprisingly, most marsh emergents are not sufficiently well adapted to water that their seeds will germinate in deep water. In fact, some either will not germinate at all or have very specific requirements governed by oxygen levels and temperature, as well as moisture. The cracked but wet and muddy bottom of a marsh is a superb seed bed for the marsh-edge plants like sedges, softstem bulrush, moisture-loving grasses, arrowhead, willow and cottonwood trees (which can be a serious problem in drawdowns), and even the deeper water plants such as hardstem bulrush and cattail (Meeks 1969, Weller and Fredrickson 1974). Such dry conditions are not

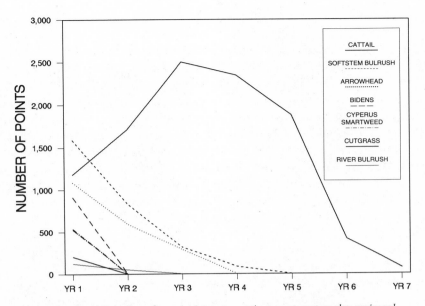

Figure 5. *Based on their incidence along point-intercept vegetation transects, wet-meadow species and mudflat species declined concurrently with increasing abundance of cattail (Weller and Fredrickson 1974). Following reflooding, wet-meadow species such as beggar's-tick (Bidens) and smartweed were eliminated in one or two years. Marsh-edge species such as arrowhead and softstem bulrush survived two to four years of flooding, whereas cattail increased for several years until it was eaten out by muskrats or floated up by high water.*

suitable for the submergents such as sago pondweed or water lilies that do germinate underwater. Thousands of seeds survive in a dormant condition on the marsh bottom and, even in the center of a marsh where such plants have not grown for five to fifteen years, buried seeds of marsh plants will suddenly germinate (Billington 1938, Crocker 1938, Goss 1924, Shearer et al. 1969, Shull 1914). Marsh ecologists term these residual seeds the "seed bank" (van der Valk and Davis 1978). Thus, the entire exposed marsh bottom may be green with small plants that at first look very much alike. As they develop at different rates, influenced by moisture or water level and other factors (Harris and Marshall 1963, Harter 1966), one finds slender, linear-leaved cattail and bulrush plants side by side with willows and sedges and other marsh-edge plants.

As water returns to the marsh, the survival, growth, and reproductive success of plants are influenced by the time of year, rate of flooding, degree of inundation, water clarity, herbivore activity, wave action, and many other forces (Kadlec 1962, Meeks 1969, Millar 1971, Weller and Fredrickson 1974). Most of these marsh-edge plants are water-tolerant enough that they survive from one to three years (Fig. 5), but the better-adapted deep-marsh plants like cattail outcompete other species in the deeper water depths (Weller and Fredrickson 1974). Gradually, the plant species composition of the wetland is set by its new average water depth as related to the germination history during drawdown and reflooding. Submergent plants probably germinate at the shallow water depths and may not become conspicuous for several years. Because different species of submergents also are adapted to different water conditions, those such as sago that favor shallow depths suffer under high-water conditions.

With higher water, marshes gradually open. Part of this opening may be due to herbivores (to be discussed next), but much vegetative loss occurs also because of the buoyancy of tubers of cattail and water lilies, which store larger quantities of food. As muskrats dig out tubers, and fish root around the bottom, even large clumps of cattail may be seen floating around the marsh, moved by the winds. Lodged plants may establish new root systems and stabilize open shores. Just as the species composition of marsh plants generally increases as one traces the history of pond-to-marsh development in any series, so the reflooding of a dry lake or marsh basin reverses that pattern over a period of three to five years. A tally of germinated plants on the mud flat might show virtually all the species common to all depths of a marsh, but these gradually are reduced to a few water-adapted species; and one or two dominants such as cattail, hardstem bulrush, or reed may be the only vegetation present after several years (Fig. 5). This decline in species richness seems unique, for in most plant communities, the tendency seems to be for variety to increase.

Muskrat Populations. Drought virtually eliminates muskrats from marshes (Errington 1963, Errington and Scott 1945). Some emigrate, as is obvious from mortality on highways; some are eaten by foxes and mink as they try to persist in submarginal habitat; and reproduction all but ceases. It does not take long for them to return, however, when water returns to the marsh. The vegetative growth is explosive (and probably highly nutritious), and muskrats may move in during summer even though the water depths are insufficient for winter. Their small numbers are hardly noticeable from the shore, and the first winter after flooding may show few lodges. But muskrats are microtine rodents like meadow mice, and their capacity for reproduction is impressive. They can have four litters per year with six or more young each time, and even their early young may produce a litter their first year when food is good and competition is reduced (Errington 1963). Exponential growth is the rule at this stage, and in several years, populations may increase manyfold (Fig. 6) (Weller and Fredrickson 1974). This explosive growth has a bearing both on the marsh system and on management of that system by humans, which we will consider later.

In the marsh system, the burst of herbivores means that plant growth may be used as rapidly as it is produced. Food resources are selected carefully, and their removal is perceptible only as one canoes through a marsh and finds entire platforms made of crisscrossed plants. The more edible parts (rootstocks, rhizomes, and tips) may be removed, and the harder and older stems are used as a platform and left to rot. But the most dramatic removal takes place in late summer, when vast areas may be cut to build lodges that must survive the winter and that the muskrats must have for protection from weather and predators. The construction of lodges in this early stage of the marsh reestablishment is an especially interesting phenomenon, for either because of reduced competition or because of the robust growth of dominant plants at this stage, lodges may be huge, where-

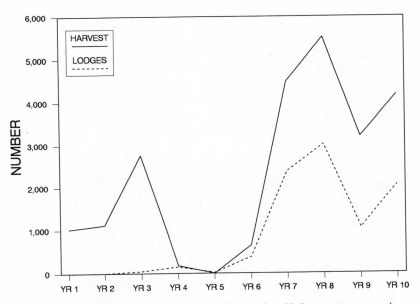

Figure 6. *Muskrat populations may explode after the reflooding of suitable dense emergent cover such as cattail. This example from Rush Lake in Iowa shows the harvest of muskrats that increased from 640 in year 6 following reflooding, to 5,509 in year 8 (after Weller and Fredrickson 1974). In this isolated marsh, few had remained after drawdowns in years 4 and 5, and, even with immigration, reproductive rate must have been phenomenal. This is a harvest of about fourteen muskrats per acre, and probably represented only 50% to 60% of the total resident population.*

as they are half that size later when populations are high and materials are less abundant. Such cutting comes so quickly that the opening of even a dense marsh can occur in a few weeks. Thereafter, the rate of opening depends on factors that influence muskrat reproduction and overwintering success. Reproduction slows when populations climb, by virtue of reduced litter size, reduced number of litters per year, and reduced survival of the young (Errington 1963).

Such muskrat populations can build to a point that they strip the marsh (Errington et al. 1963, Weller and Spatcher 1965). Dependent on the basin

Cattail seeds by the tens of thousands drift from each head. Germination of seeds may be very high on suitable mud flats, but cattail also spreads continuously by vegetative propagation.

Revegetation of Rush Lake near Ayeshire, Iowa, takes place gradually after a muskrat eat-out. Partial (top) and complete (middle) drawdown resulted in germination of dormant seeds that had been underwater, and produced an excellent stand of emergent vegetation attractive to a variety of birds (bottom). Openings of this size and variety rarely occur during the first year, but the wetland usually is not attractive to birds until they develop.

A muskrat feeding platform is built of leafy remains of cattail after the tuberous base has been eaten.

shape, this stripping usually occurs in the center first, because muskrats seek the deeper depths where they are less vulnerable. If water levels rise, muskrats move toward the shore and strip even that. Sometimes they leave the marsh to feed in wet meadows or cornfields, leaving little food anywhere. A population crash is inevitable. If, on the other hand, water levels decline, the vegetation may be saved, but the muskrats must move out of the marsh for lack of suitable overwintering depth. If water levels decline during winter, muskrats are forced to move at a vulnerable time, and high mortality results (Errington and Scott 1945). Trapping can accomplish the same thing on small areas, by greatly reducing a muskrat population and preventing the stripping of a marsh, and it is an important management tool.

Impact of Vegetation Changes on Bird Populations. From the general descriptions of both the kinds of habitats various birds use and their species-specific adaptability, it is obvious that some bird species respond to slight changes in habitat. Vegetation spatial patterns and vertical structure seem to be the major visual clue to which marsh birds respond initially. Undoubtedly, food must be present next, and at a density that minimizes time wasted in foraging. For breeding birds, nest sites must be present, and

One view (top) of Goose or Anderson Lake near Jewell, Iowa, shows the dense emergent vegetation that occurred in summer of 1959; another (bottom) shows the lakelike conditioning following a muskrat eat-out in 1962.

our data on the usual sites selected gives a quantitative measure of what is used (but it is difficult to relate to what is not used). We gain further insights into preference as we observe changes from year to year. Thus, yellow-headed blackbirds that nest over water may return to their breeding marsh of a previous year to find that muskrats have stripped much of last year's cover. The marsh is relatively unattractive to the birds unless emergent vegetation is standing in water. Some birds may battle for the few remaining sites, but others abandon them to the redwings, which tolerate lower, less robust, and more open marginal vegetation. Even the redwing males may find the habitat less attractive to females and eventually abandon their territories (Weller and Spatcher 1965).

Nest-site selection of Forster's and black terns shows even more specific association with a microhabitat. In general, black terns will use either very small marshes or small openings in large marshes. Their manner of flight and their adeptness at feeding on invertebrates at the water's surface make this behavior practical. Forster's terns are fish feeders, however, and generally are associated with larger marshes or lake edges or very large pools in larger marshes. But nest-site selection is rather species-specific, with Forster's terns using large, dry muskrat lodges (or other dry sites) whereas black terns select low, deteriorating lodges (or build soggy sites of wet vegetation). Thus, black terns may move in immediately after flooding, using sites created by flooding of sedges and other low, wet-meadow plants. Forster's terns are prevalent in the midstages of the reflooding, when large lodges have been built but only after large pools of open water are present. Hence, habitat change creates a shift not only in spatial distribution but also in species richness and numbers of birds (Fig. 7) (Weller and Fredrickson 1974). Figure 8 summarizes these patterns of change (Weller and Spatcher 1965). After reestablishment of dense stands of emergents following drawdowns, muskrats slowly invade and build lodges, creating small openings. As further vegetation is cut or eaten, and these pools increase in size, the area becomes most attractive to the greatest variety and the largest numbers of birds.

The extreme cases of habitat change do bring about the elimination of some species. When bottom soils are dry and weed-covered, pheasants and meadowlarks move in, and even mourning doves nest on the ground (Weller and Spatcher 1965). Song sparrows and other upland passerines may nest in low shrubs or trees. When water levels are too low, waterfowl simply do not stop (Krapu et al. 1970), and some move long distances to nest (Smith 1970); other species stop but do not nest (Rogers 1959). When

A. Oak-rimmed Rush Lake in northwest Iowa is at a dense-marsh stage following drawdown and revegetation.

B. Bogs are usually acid lakes or pools with an overgrowing mat of mosses, sedges, cattail, shrubs, and even trees. Marsh vegetation occurs in these settings, but such wetlands usually are classified as bogs or bog lakes rather than marshes.

C. Drainage and filling are gradually eliminating extensive wetland complexes in typical prairie potholes near Roseneath, Manitoba (Kiel et al. 1972).

D. Arrowhead or duck potato is a common wet-soil or shallow-marsh plant that produces rich tubers eaten by ducks and muskrats.

E. A natural wetland overfertilized by livestock wastes is covered with a resultant bloom of floating duckweed. The effects of grazing on vegetation are evident.

F. Hardstem bulrush is among the most water tolerant of the emergent plants, persisting in shallow lakes as well as the deeper sections of marshes. However, because of its effective root system, it can be equally tolerant of drought.

G. Yellow water lily is a floating-leaf plant common to deep marshes and shallow lakes. These leaves, like branches of a spreading tree, tap sunlight over a large area and store nutrients in huge rootstocks that supply the initial thrust of the next season's growth.

H. A pair of Forster's terns settles at a nest on a muskrat house in a stand of cattail. Nest success usually is quite high at such sites.

I. A female least bittern on the nest shows the ""freeze" position for which bitterns are so well known. This inconspicuous bird can be extremely common but will remain unkn to passersby who do not penetrate the cattail and bulrush of the deep marsh

J. In the winter prairie marsh muskrat lo protrude through the snow and ice. Most birds are gone, but muskrats are active underwater, and mink and weasels are active all year.

K. Fire in marshes is dangerous during the breeding period because of the losses to nesting waterbirds. However, winter burns may be useful in opening dense vegetation, recharging the nutrient base, and stimulating certain types of vegetation.

L. This Waterfowl Production Area is located in the more rugged North Dakota coteau, or hilly uplands. Such WPAs are owned or leased by the U.S. Fish and Wildlife Service and managed for wildlife production. Hunting normally is permitted, but disturbance should be minimized during the breeding season. This is a semipermanent wetland in the open phase, showing diverse emergent plants along the edge and sparse hardstem bulrush "islands" in the central marsh. Water depths typically range from one and a half to three and a half feet.

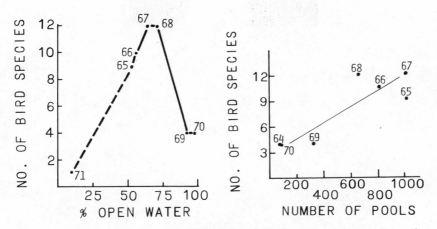

Figure 7. *Bird species richness (the number of different species) is related to the area (%) of open water and the interspersion of open water in dense cover. These examples suggest that the highest number of species occurred with 50% to 75% open water and the maximal number of small pools in dense emergents (Weller and Fredrickson 1974). The figures may vary from area to area dependent on plant life form, species composition, and nearness of other wetlands, but structural features of this type are clearly important in attracting birds.*

■ WATER ▨ CATTAIL ▦ HARDSTEM			
WATER DEPTH	SHALLOW	MEDIUM	DEEP
VEGETATION	DENSE	MODERATE	SPARSE
BIRD POPULATIONS	NUMEROUS INDIV.	MANY INDIV.	FEW INDIV.
BIRD SPECIES RICHNESS	FEW KINDS	MANY KINDS	FEW KINDS
MUSKRATS	FEW	MANY	FEW

Figure 8. *As a marsh passes from dense vegetation to open water because of the action of high water and muskrat activity, considerable change takes place in the numbers of muskrats and birds, and a major change in bird species richness. The same pattern of animal numbers and species richness is characteristic of wetlands that remain in this condition for long periods owing to water regimes.*

After reestablishment of dense stands of emergents following drawdowns, muskrats slowly invade and build lodges, creating small openings. As further vegetation is cut or eaten, and these pools increase in size, the area becomes most attractive to the greatest variety and the largest numbers of birds.

reflooded, the same vegetation might be used by a variety of semiaquatic species of birds before the development of other, more aquatic, plants. In such situations, yellow-headed blackbirds, coots, or canvasbacks may nest in willows when they are flooded.

Some adaptable species such as coots and pied-billed grebes are tolerant of a wide range of vegetation, as long as foods and nesting material are present and water depth is adequate for swimming. Both species require some small openings in dense vegetation, and favor semiopen conditions, but are tolerant of rather open situations in which their nests are totally exposed— as long as they have nest material. Emergent cover as such is not necessary, but their nests must be protected by old stubs of emergent plants or, occasionally, dense beds of submergents near the surface that keep wave action from destroying the nests. Only wide-open lake conditions eventually discourage breeding birds from returning. Eared grebe colonies often occur in wide-open but shallow lakes, where the birds build nests in and of submergents that are at water level and that protect the nests from wind damage.

Figure 9 diagrams these general influences on changes in bird species and populations from dry to wet cycles, and from dense to open vegetation.

The Importance of Food Resources

Birds and other wildlife species feed heavily on insect and other animal matter, which is most abundant during the summer (Fig. 10) (Bergman et al. 1977, Danell and Sjoberg 1977, Hohman 1977, Swanson and Meyer 1973). All too little has been written about how invertebrates (Schroeder 1973), frogs, and other food resources in marshes are related to vegetative zone or marsh type. Hence, it is more difficult to assess the effect of water-level changes and vegetation changes on these animals. One of my former graduate students gathered data on several wetland types in the same region, and formulated a simple model of what would happen with change in marsh type from newly germinated areas to those opened by high water

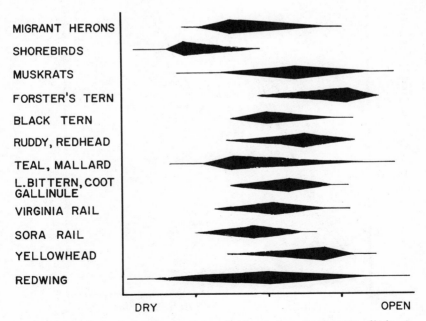

MIGRANT HERONS
SHOREBIRDS
MUSKRATS
FORSTER'S TERN
BLACK TERN
RUDDY, REDHEAD
TEAL, MALLARD
L.BITTERN, COOT GALLINULE
VIRGINIA RAIL
SORA RAIL
YELLOWHEAD
REDWING

DRY   OPEN

Figure 9. *This schematic diagram shows the influence of habitat selection on the composition of bird species of a wetland (Weller and Spatcher 1965). Some species such as shorebirds and waders occur only during the drier or moist-soil stages, whereas redhead and ruddy ducks favor the open or deep-water range.*

and muskrats (Fig. 11) (Voigts 1976). This model suggested that major changes in invertebrate species and numbers do occur and that they could have a significant influence on birds that use the area. For example, plankton are abundant when vegetation is decomposing and nutrients are being returned to the aquatic system. Shovelers have bills highly specialized for straining plankton, a food resource commonly used by fish and other aquatic animals. It is not unusual to see a pair of shovelers on any small wetland, but large flocks (in spring migration or after breeding) are common on open marshes where few other ducks occur and where foods would seem scarce. They are there for the plankton.

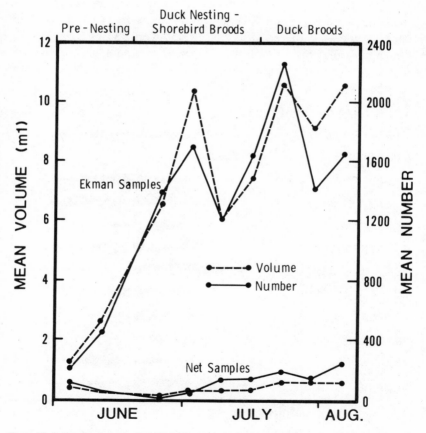

Figure 10. Seasonal variation in the bottom (Ekman dredge samples) and the planktonic (net) invertebrates of some Alaskan tundra ponds show maximum numbers and volume in July, when duck and shorebird numbers are at their peak (Bergman et al. 1977).

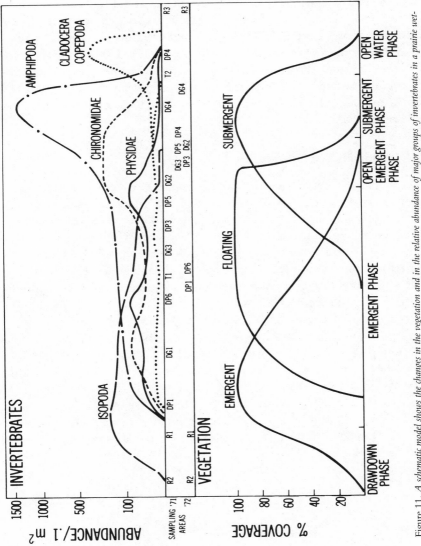

Figure 11. A schematic model shows the changes in the vegetation and in the relative abundance of major groups of invertebrates in a prairie wetland. This figure was constructed by Voigts (1976), based on pooling of observations of populations and conditions in several different wetlands. Its value may not be so much its precise accuracy for any single wetland as its demonstration of the dynamics of the vegetation and invertebrates.

Diving ducks dependent on snails or on bottom-dwelling midge larvae also may occur in open-marsh types, but other marsh stages are more productive of even these organisms and, therefore, of these ducks. Dabbling ducks such as blue-winged teal that feed on snails and other invertebrates in more shallow, densely vegetated areas shift to larger wetlands during drawdown phases in open marshes, showing both flexibility and adaptability to rich resources (Swanson and Meyer 1977). Extensive data from observational and experimental studies suggest strongly that these invertebrate resources are the major influence on waterbird use and production in marshes (Bengtson 1971, Eriksson 1978, Joyner 1980, Kaminski and Prince 1981).

The Importance of Instability

Most current evidence shows that stabilizing water at high levels is detrimental to marshes because it tends to create lakelike situations, where production would favor aquatic rather than wetland species. Organic production in emergent plants is reduced, as deep water favors submergents or floating-leaf plants. Invertebrate production probably is less, especially of detritivores, and total nutrient availability probably is lower. Toxic substances may build up under the anaerobic conditions of continuous flooding (Cook and Powers 1958), and the return of nitrogen and phosphorus to the system may be slowed (Kadlec 1979). At low levels, marshes become too dense and more like terrestrial systems. But periodic drying and reflooding is generally beneficial. This need for seasonal instability, though, should not be interpreted as a need for erratic water-level changes at any time of the year. Fluctuations that are too rapid may cause mortality of muskrats (Choate 1972) and ducks (Hochbaum 1944) if such a change occurs during the breeding period. Some mortality is expected, and further research is needed to evaluate the significance of long-term gains versus short-term losses.

CHAPTER 8
Management and Restoration

Philosophical Considerations

To manage or not to manage is a controversial issue in resource conservation, stemming from a new public awareness of environmental issues and activities. The opinions are, unfortunately, strongly divided, but the reasons for the philosophical difference seem at times unrelated to the wildlife resource. Rather than review the arguments on either side of this issue, it seems more profitable here to clarify some goals in wetland conservation and then to discuss methods for achieving them.

In view of the extensive loss of wetlands in North America and around the world, our major emphasis must be to conserve marshes as communities in a natural state for present and future generations. In many cases, this policy will mean only acquisition and protection (a management decision), but, in some cases, human modification or manipulation will be valuable, if not essential. I suggest that our national policy include the preservation of wetland complexes in as typical a natural state as possible. This action might involve purchase, in a balanced pattern, of representatives of all typical wetland types in an area, or, where this is impossible, management of some more abundant types to stages comparable to those lost. This procedure will help maintain faunal and floral diversity typical of that region and avoid endangering some species while overproducing others. Moreover, preservation will provide numerous other natural functions that wetlands perform in the area. This approach is, of course, more problematic than the preservation of a single wetland or the management of that unit; it requires a recognition that wetlands function as a complex for more mobile species such as birds (Weller 1979), and that the modification of one wetland may influence use of others nearby.

A general philosophy of management might be to "leave well enough alone" when a wetland is relatively undisturbed and seems productive, but to use natural techniques when human-influenced wetlands are not productive. Moreover, when management is deemed necessary, we must understand that a marsh is a complex system and must be managed as such. Management goals must consider all species, not single ones or small groups, because most changes affect all organisms in the ecosystem.

To reduce excessive, often unnecessary, and overly artificial management programs, goals should emphasize (1) long-term productivity, under the most natural conditions possible, by maintaining natural values through the use of natural processes, (2) long-term over short-term benefits, and (3) conservation programs that serve the greatest range of public interests (Weller 1978a). Conservation agencies, however, may take different attitudes toward management needs and techniques, dependent on whether they focus on maintenance of wildlife diversity or on selected harvestable species favored by a user clientele (Bishop et al. 1979). Both should be considered concurrently, and there are many new programs favoring watchable wildlife and less well-known species (Hands et al. 1991).

Situations that may encourage management are those extreme stages of succession that seem to cause many species of wildlife to reduce their use of marshes. For example, excessive water from modified drainage systems creates lakelike environments. Because revegetation is essentially restricted to drawdown conditions, several methods may be used to achieve the germination phase, but those that have their basis in natural events are preferred. Inflow modifications as well as outlet structures may be necessary. Increased eutrophication, at either open or dense stages, could be a product of pollution with excessive nutrients. Management to correct the problem might involve upstream impoundment or changes in watershed cover plants, but too little is known as yet, either of the problem or its solution.

The opposite extreme is vegetation so dense that birds are not attracted. This condition is more difficult to assess and to control, but even songbirds seem to have a preference for patchy as opposed to continuous cover; open water seems to be a vital ingredient (Weller and Spatcher 1965). The opening of this dense cover adds a diversity of plant life and, presumably, invertebrates, and encourages bird use. In large marshes, we have found that somewhere near a fifty-fifty cover-to-water ratio is ideal (Weller and Fredrickson 1974), but diversity also can come from several nearby marshes of a complex that are in different successional stages. Regardless of the specific goal, several options are available for creating this wetland diversity dedicated to enhancement of wildlife diversity as well as other functions in the region.

Acquisition of Wetlands

The purchase of marshes and other wetlands as wildlife refuges is well known, having been the only feasible way of protecting unique areas for bird-nesting colonies or as strategic migration stops. The U.S. Fish and Wildlife Service, Ducks Unlimited, the National Audubon Society, the Nature Conservancy, and numerous state and local conservation agencies have thus acquired significant acreages of marshes in refuges, sanctuaries, and preserves.

In addition to the National Refuge System acquired mainly in the 1930s and 1940s, the Fish and Wildlife Service started in the early 1960s to lease or purchase small wetlands known as Waterfowl Production Areas (WPAs) to help maintain continental production of waterfowl and other migratory birds. In North Dakota, nearly 1 million acres have been purchased or leased as of 1980 (Madson 1980). The WPA program ran into serious difficulty, however, because of local and state opposition to taking land out of the tax base, to increasing the amount of federally owned land and federal influence in state issues, and to increasing limitations on deeds that would restrict property resale. In addition to the special goal of buying wetlands, these funds have also been directed toward the purchase of some beautiful wet prairie and uplands in the Prairie Pothole Region that will serve wider audiences than duck hunters. In addition, the Nature Conservancy and other private foundations have conserved wetlands with the intent of preserving unique, representative, natural communities as part of our national heritage. Their acquisitions tend to preserve all forms of typical wetlands, regardless of their wildlife values. In addition, such organizations often perform a unique service in helping government agencies in purchases since private groups can act more quickly and can hold property until other action can be taken.

More recently, these conservation agencies and foundations have banded together in an international effort termed the North American Water-fowl Management Plan to acquire both breeding and wintering habitat under careful management. Small but important habitats are being preserved as a result of land abandonment or through the efforts of organizations and individuals with special interests in wetland wildlife. Moreover, the benefits go far beyond waterfowl or wildlife, to include flood retardation, water purification, green space, and passive recreation.

Natural Methods: Water–Level Regulation

Using natural processes in management is most likely to simulate natural events, conditions, and responses (Anderson and Glover 1967). Techniques based on such processes are the least expensive to use and have the greatest permanence. As mentioned earlier, the dominant forces influencing vegetation dynamics are hydroperiod, seasonality, water depth, water-level fluctuation patterns, herbivore utilization of plant materials, ice action, and possibly fire and nutrient cycling.

Water-level management is generally only feasible when a water-control structure has been constructed either to restore a marsh or to aid in water-level stability. Whether such structures should be added to a natural marsh to provide water-level control is a debatable question for which there can be no universal answer. In some cases, the success of management may justify the effort. In others, the area may require it because of such changes as modified stream flow, lowered water tables, increased sedimentation due to wind or water erosion, and increased disturbance or consumption of vegetation because of artificially high levels of wildlife or domestic livestock. Massive dams usually are unnecessary, excessively expensive, and no more functional than smaller ones (Atlantic Flyway Council 1972, Weller et al. 1991). Low earthern structures can be built that differ little from natural "glacial walls" but that effectively impede drainage.

Conflict has been common between biologists, who want to design management systems to maximize biological productivity, and engineers, who wish to devise water-management systems. Communication between the two groups often has been poor or lacking, and the outcome sometimes disastrous for wildlife. Engineers can design suitable water-management systems for marsh management only if they first understand the biological goals of management, and the sensitivity of the marsh community to water fluctuations. Overengineered structures tend to lack the sophisticated controls essential for good marsh management. A common failing is a control structure that cannot regulate levels within the necessary precision of one or two inches. Another weakness is the siting of the control structure in such a way that complete drainage of the basin, an essential device for creating suitable plant seedbeds through oxidation and decomposition or for eliminating carp or muskrats, is impossible.

Water-level control through pumping out water or pumping in water has been exercised commonly, but it is expensive in equipment, time, and operational costs. However, the process and the results are natural; the decision to use this procedure is a judgment that must be made in light of other possible options or potential pressures for misuse of the system.

Regardless of how water-level regulation is accomplished, it may be used to flood out dense or weedy vegetation established during dry periods, or to dry out the marsh for revegetation. Drawdowns for revegetation may be (A) complete, when major restoration is essential in an open marsh; or (B) partial, when vegetation needs to be encouraged or herbivores discouraged.

A. Complete drawdowns are common when all central vegetation has been lost owing to high water levels, muskrat eat-outs, winter kill, and occasionally plant disease. Procedures are as follows (Linde 1969, Weller 1978a):

1. Lowering water levels during the wetland plant growing season permits aeration to enhance decomposition of bottom vegetation, as well as germination of naturally occurring seed and recovery of established but flood-stressed emergents and even submergents.
2. Timing of the drawdown in the growing season is a crucial locally based decision, because too early a drawdown may induce emergents and willows that become weedy and too late a drawdown may favor valuable but annual species that do not tolerate much flooding.
3. The length of drawdown probably is a function of the recentness of the previous drawdown, but drying of the soil and breakdown of vegetation to release bound nutrients may require most of the growing season. Overwinter drawdowns have proved effective. Late fall or early winter (posthunting) drawdowns may not produce germination in the fall but can be left moist until late in the next summer (prehunting), so that duck hunting is not eliminated. Muskrats and carp will be drastically reduced over the winter.

4. Reflooding should be a gradual process, to avoid flotation of emergents, direct scouring of other plants, or plant mortality due to the turbidity of muddy waters.

5. Water levels should be regulated mainly for vegetation growth, diversity, and survival during the second (first reflooded) season, as long-term trends demonstrate a gradual decline of emergent vegetation with stable or high water levels (Fig. 12). Some concern for wildlife must be deferred at this time, since long-term production of wildlife will be enhanced in later seasons.

6. Subsequently, marsh management for the benefit of wildlife will consider welfare of the vegetation. Knowledge of species requirements is essential, but the "system" will be self-forming and dynamic.

7. Some submergents may germinate on mud flats, but most germinate underwater because they are better adapted to aquatic conditions. Excessive depths, however, especially of turbid water, are detrimental to submergents.

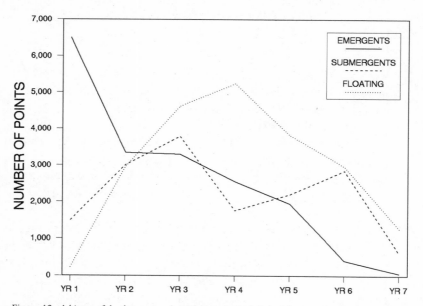

Figure 12. *A history of the three major plant life forms in Rush Lake following reflooding (Weller and Fredrickson 1974) is shown. Whereas floating or submergent plants varied in number according to water levels and wave action, emergents gradually declined, as a result of muskrat use for lodges and food.*

8. After several years, reduction of herbivores by late fall drawdown may conserve vegetation. Not all muskrat populations "explode," but many do, and lowered water level increases their vulnerability to traps.

B. Partial drawdowns should be used where vegetation is seriously reduced, wildlife use has declined, or water levels have stressed vegetation. Procedures include:

1. Water levels should be reduced to meadowlike depths, to encourage vegetative propagation of emergents and germination and growth of submergents in early summer. Wildlife use by species such as inland diving ducks and coots that favor deep water will decline; use by waders, shorebirds, and dabbling ducks may increase markedly.
2. This low level should be retained or even lowered farther in late summer, and returned to near-normal levels in early fall after plants are less vulnerable.
3. It is best not to keep freeze-proof depths, except where plant density is high and muskrats are to be encouraged. The presence of the non-native carp is a consideration, because this fish is usually deemed an undesirable species in marshes.
4. As vegetation recovers, levels are regulated to allow nesting and plant consumption as desired.

Natural Methods: Herbivores and Other Vegetation Management

Management of muskrats, nutrias, beaver (Buech 1985), or other herbivores is a valuable but more difficult management tool, since it depends on the efficiency of population regulation (Cartwright 1946, Krummes 1941). If management guidelines are flexible, trappers can effectively control muskrats and nutria where water control is lacking or limited. In most midwestern marshes, trapping too late in the population boom creates more problems than does overtrapping the population. Not only are muskrats difficult to trap, but unpredictable weather conditions, low fur prices that discourage trapping, and restrictions on places and methods of trapping reduce trapping success, make population control difficult, and are detrimental to results of marsh management.

The use of livestock grazing to manipulate vegetation has met with mixed reaction by wildlife managers, and has been applied less commonly to marshes than to other types of cover. In many cases studied to date, extreme overgrazing has had negative effects on duck nesting and habitat quality in uplands and marshes (Weller et al. 1958), and even moderate grazing has been detrimental. Others studies of well-managed pastures have had more positive results. We must maintain an open attitude on the use of grazing as a tool in marshes until more careful experimental data are available. After all, bison grazed the entire West and Midwest and must have exploited marsh resources at times.

Fire has been used extensively in Gulf Coast marshes (Chabreck 1976, Hoffpauir 1968, Lynch 1941) but less commonly in glacial potholes. Natural fires must have occurred regularly before the intervention of humans, and possibly functioned, as do drawdowns, to eliminate the bulk of plant biomass and to return some major nutrients to the system. In fact, this may be a danger; fires to control cattail have in some cases enhanced the problem because available nutrients, without water-level control to flood the burned stalks, encouraged even more growth (Weller 1975a). A great difficulty in using fire is that it often spreads too far, too fast. Certainly, fire can never be resorted to in a marsh during the breeding season, as disastrous results could occur when early-nesting ducks explore the tinder-dry marsh vegetation of the previous year's growth (Cartwright 1942). Some positive benefits of burning reeds have been reported for managing marsh-edge nesting areas for dabbling ducks (Ward 1942). Fire has been used to deepen bog wetlands in some north-central states, but it is very difficult to direct and to extinguish (Linde 1969).

The release of nutrients by a fire has caused some speculation and some experimentation on the use of fertilizers to enrich marshes. This practice seems wasteful, since marshes seem to be nutrient sinks anyway, but some positive short-term responses have been noted.

Artificial Methods

What I have termed "artificial" simply implies high-energy or high-cost input to achieve goals quickly. Such actions as seeding and planting of vegetation often were attempted during early marsh development for the protection of shorelines or for production of waterfowl food. Failures were common, and such efforts now are restricted to special cases of intensive management of emergents or the large-scale production of waterfowl foods on migration stops or wintering areas. In most cases, natural seed banks seem more than adequate if water levels can be manipulated to induce germination.

Basin deepening is the most common approach to opening up dense marsh, usually involving dredges, draglines, bulldozers, and blasting (Linde 1969). Various of these methods have been shown to produce abrupt edges less suitable for emergent plants and, therefore, less attractive to swimming waterbirds. Moreover, they rarely achieve the fifty-fifty cover-to-water ratio that seems ideal for most waterbirds (Weller and Fredrickson 1974). I have seen some examples where such methods have produced openings that, after a few years, appear natural (Strohmeyer and Fredrickson 1967). In these cases, blasting or bulldozers were used rather than a dragline in a fairly dry situation where grading the shoreline was possible by either judicious placement of explosives or the bulldozer scraper blade (Provost 1948). Creation of isolated wetlands has proven effective in some situations, with the larger, less symmetrical, and most clustered ones proving most attractive to waterfowl (Leschisin et al. 1992). Blasting small wetlands with fertilizer (ammonium nitrate) triggered by dynamite tended to produce "holes" rather than "basins" with gradual slopes until packaging allowed charges of different sizes. Draglines, generally used in very wet situations, also are less easily controlled, especially underwater. An especially drastic method that uses a dragline is level ditching (Mathiak and Linde 1956). Long troughs are dug across a nearly dry marsh basin, and the spoil is deposited as flattened ridges between these canals. Evaluations of such efforts generally have shown positive results for muskrat and waterfowl populations and acceptable cost-benefit ratios, but the product is esthetically dubious. What this system achieves that other efforts lack is a good balance between cover and water, since whole basins rather than portions tend to be modified during the operation. If this same interspersion could be achieved with a more natural basin formation, the same benefits should accrue, as well as a more esthetically pleasing outcome, probably at lower cost. However, the technique has proved useful in encouraging the preservation of wetlands for fur production and harvest, and is preferred to the drainage of shallow wetlands less productive of furbearers.

Island building with some of the same tools also has been done, especially for creating nest sites protected from predators; such islands also enhance cover diversity and increase edge. However, naturalness of the island edge is difficult to achieve and, as a result, such islands have been more successful for upland nesting waterfowl, such as the Canada goose and gadwall, than for birds that nest at the marsh edge, such as coots and diving ducks.

Artificial methods also may involve short-term programs to modify emergent growth, such as cutting of cattail on ice in winter when water levels are low. When reflooded in spring, plant growth is inhibited, and the area may remain open for several years. Although it is costly, this system at least does not produce permanent damage (Weller 1975a).

Some herbicides have been used successfully to create open areas in marshes, but they are difficult to apply and to control. Moreover, the robust nature of many of the target emergents is such that they may stand for several years before an opening results that attracts birds. Moreover, few of the herbicides are so species-specific that they do not eliminate more favorable species. With several introduced nuisance plants such as water-hyacinth, hydrilla, and purple loosestrife, however, wetlands are impossible to keep open without herbicides.

In addition to the methods applied to manage the marsh itself for wildlife, numerous programs have been designed to provide completely artificial habitat features for nesting birds. These include artificial nest sites for hole-nesting ducks such as wood ducks (Bellrose 1990) and for platform-nesting birds such as mallards or Canada geese (Bishop and Barratt

Mallard nests sometimes are built on muskrat houses. Nest success in these situations is quite high because of freedom from predators on this muskrat-built "island." Such behavior preadapted the species for use of nesting platforms built by people.

1970). They reach their extreme of artificiality in washtubs on posts and in a fiberglass nesting structure with a platform for mallards or Canada geese above and a hole for wood ducks below! Most of these devices, though, have gained acceptance by birds, and nest success often is high. Whether we can afford to produce, maintain, and replace a significant number of such structures is one issue; the esthetic element is another consideration.

Fixed or floating platforms have been used to attract loafing or nesting ducks and geese (Brenner and Mondok 1979, Sugden and Benson 1970). These structures should be used with discretion, and efforts should be made to have them appear natural.

Such methods and many others advocated may have their place in newly created, artificial, or highly modified situations. Their use as a general operational policy needs to be examined economically and esthetically. They are costly in terms of people-power, dollars, equipment, and time, and often leave esthetically displeasing results.

New Marshes from Old: Marsh Restoration

Fly over the prairies in a wet spring and all low areas—farmed or not—will show the basins of former potholes. As the water seeps through the drainage tiles and ditches, these areas usually, but not always, become dry enough to farm. Some never make good farmland, because the soil is always too wet or too peaty for good crops. In such cases, state and federal agencies have tried to reestablish marshes, and fortunately, it seems not too difficult to do in the Prairie Pothole Region.

In those wet years when flooded fields do not drain, marsh emergents often germinate from long-buried seeds, surprising everyone. Such plants may grow rapidly and densely, and their growth creates the key physical structure of a marsh, attractive to birds that may respond during the very next nesting season (Sewall and Higgins 1991). But obviously, it takes years to duplicate the complex flora and fauna one expects in a long-established, natural marsh. Birds that respond to the life form of the vegetation and the water recognize and test new marshes, but presumably only those where suitable foods have developed can hold a population for long.

Because so many plants and invertebrates of marshes survive as a result of tolerant seeds or eggs, and because these organisms are distributed by wind, birds, and probably other animals (Atkinson 1971, DeVlaming and Proctor 1968, Malone 1965), diversity seems to come quickly to marshes.

Much research is needed in this area because more of this type of work will be necessary in the future.

Marshes Where They Were Not

The creation of entirely new marshes once was only the incidental result of other activity. For example, some functional but not necessarily attractive marshes have been created by the borrow pits of highway or other construction, and by mining for gravel (Catchpole and Tydeman 1975, Harrison 1970, Toburen 1974). Coal mining normally results in highly acid water, with reduced plant and animal life, but new methods of handling topsoil have reduced this problem in most areas. However, lakes or ponds or marshes are no longer the end product of coal mining because laws often require a return to premined conditions or to agricultural land, making these areas unsuitable for aquatic wildlife. With proper planning and effort, and little increased cost, such construction digging could produce natural-looking and effective marshes by creating basins of irregular shapes and by modification of steep sidewalls to gradually sloping shorelines.

Recently, mitigation—legally mandated compensation for damages to wetlands through human activities—has produced many experimental efforts at marsh creation. The loss of wetlands due to highway construction has, in some states, been mitigated by creation of artificial wetlands in contiguous roadside ditches (Blair and Sather-Blair 1979). Such efforts cost thousands of dollars per acre of marsh created, but the expense has been tolerated because of the economic gains resulting from such developments.

We can define wetland creation as the difficult and often impossible act of creating a new wetland on a site that was formerly terrestrial, or from a lake or bay flooded so deeply that marsh seed banks no longer occur (Pederson and van der Valk 1984). Rarely do we test for a seed bank, but if we select a low site that has been drained, return the water supply, and find that a variety of wetland plants germinate, this reflects a former wetland and should be called restoration. If we are engaged in a management effort in which we are trying to improve an existing wetland, it can be called enhancement and is a form of management or manipulation. Wildlife managers often used the term development for any habitat improvement action that might involve any or all of the actions described above.

In cases where damage to part or all of a wetland occurs as a result of

U.S. Army Corps of Engineers' "404-permitted" development, mitigation agreements between the participating agencies may demand one of these three actions. If the entire wetland is lost, restoration is most logical and reliable if an earlier modified wetland exists nearby. If only part of the wetland is destroyed, or its functions impaired, enhancement may be the logical option, but measuring the success of such efforts is difficult and costly. Otherwise, creation may be necessary to fulfill our generally accepted policy of no net loss of wetlands. In all these undertakings, the steps of establishing precise goals, conducting restoration or creation efforts with ecological perception and soundness, monitoring the effort, and then correcting existing problems often are poorly carried out. A careful plan of action and review should be written into mitigation agreements.

The techniques for performing wetland restoration, enhancement, or creation differ little from techniques commonly used in wildlife management (Weller 1989b), and are summarized in Appendix B. Several other books treating other specialized interests and techniques are available and included in the References. Mitigation-driven management efforts differ from waterfowl management efforts in one key way: Costs have been less of an issue, and prompt response has been mandated, with the result that planting has been used rather than seeding or natural seed banks. Moreover, engineering devices may be more elaborate and costly with the purpose of speeding the process. Such approaches may be shortsighted because the natural processes that maintain self-perpetuating systems probably are the least expensive and most effective way to accomplish the goal. It may take a bit more time, but the end product may more nearly approach the natural combination of species that allows the system to function. If, for example, one cannot recreate the natural hydrology/hydroperiod of a wetland type in an area, costs of providing water of the right type at the correct time could be enormous, and the chances of success are extremely low. Instead, the project should be abandoned at that site, and wetland restoration elsewhere considered as a mitigation alternative.

Restoration efforts, based on some of the natural processes such as succession and management strategies outlined earlier, have become commonplace in the Prairie Pothole Region where annual-plant seed banks are common, long-lived, and respond rapidly. Evaluations of these show rapid invertebrate response as well as plant germination and subsequent use by waterfowl and other waterbirds. Other wetland types characterized by long-lived perennials and special soil or water conditions (wooded swamps, playas, or acid bogs) are difficult to restore, and perhaps impossible to create from scratch. Thus, we should not become complacent about wetland loss and damage because wetlands can be replaced; such loss and damage are very serious, and the original often is irreplaceable. Moreover, there is a tendency to eliminate the small and temporary wetlands under the false assumption that the larger and wetter are better.

Birds versus Fish

It is obvious to the reader by now that a marsh is not a lake, at least most of the time. Even when it is open and lakelike, it often is too shallow to support a major sport-fish population. Yet in some areas, such marshes are excellent fishing areas for bullheads, panfish, and occasionally bass, and fishing enthusiasts take a dedicated interest in their condition. But other people are interested in marshes for their potential in aquaculture. Let us examine some of these situations and the conflicts they produce.

First, because lakelike conditions are best for fish, fisherfolk prefer high water levels, usually object to drawdowns (not realizing that these ultimately also benefit fish production), and prefer to eliminate "water weeds"—because they foul motors and fishing tackle. Several guides on weed control are designed to improve the fishery values of shallow lakes (Lueschow 1972) or the waterfowl habitat (Martin et al. 1957). A great variety of herbicides has been used, and underwater cutting bars and grinders are in production and in use. Obviously, all these systems are fighting the natural forces that lead to the plant growth: shallow water and high nutrient availability. Water-level increase, which could reduce the aquatic plants, might also make the marsh a lake, thereby destroying much of its potential for waterbirds. But the pressure for fishing areas is great, and the preference for having them in one's "backyard" is nearly as great. Hence, a major policy decision is required. This decision might best be based on the natural history of the area, not the absence of fishing "holes" locally. Those who would find the fishing exciting must pursue their quarry with canoe rather than motorboat, and suffer the conflict between hooks and water plants.

Carp have been raised in Eurasia for centuries, resulting in major protein production, and the raising of fish in ponds for commercial purposes is not new even in North America. Putting fish in ponds for recapture has been done for fifty years or more, but the emphasis has been on recovery of hatchery or other special fish stocks for release elsewhere. The use of colder northern ponds and marshes has attracted trout raisers who can put the trout in food-rich ponds and retrieve them before freeze-up (Miller and Thomas 1956). Obviously, the potential for competition between these introduced fish and native fish, amphibians, and birds is very real (Sugden 1978). Unfortunately, the competition has not been evaluated before such programs have been put into operation; this evaluation is essential.

The continuing construction of reservoirs has been responsible for extensive losses of floodplain and riparian wetlands that included considerable backwater marsh, as well as other types of wetland vegetation. These losses have been replaced with wetlands of another type, shallow lakes, and are tallied in wetland inventories as a net increase in total area (Frayer et al. 1983). However, they are of a different class and quality, being better for fish perhaps but less valuable for waterfowl and other waterbirds. They are used by certain ducks and geese as resting and roosting areas on migration and during wintering, but they reflect a net loss in wildlife habitat. Without doubt, some reservoirs have produced excellent fringing marsh, but most are too erratic in water levels to induce the best growth and development of marsh vegetation.

The Coming of the Carp and Some Alien Plants

The introduction and spread of carp in North America has produced serious and recurring problems in maintenance of marsh quality (King and Hunt 1967, Robel 1961). I cite it here as an example of human misunderstanding of biological phenomena as well as a problem in management. Carp thrive in places and at densities that no native fish seem to tolerate (Cahn 1929). As an omnivore, carp can ingest fine plant foods but can take bottom invertebrates as well. Marshes have proved excellent places for them, and muddy waters, uprooted plants (Threinen and Helm 1954, Tryon 1954), and food competition result. In northern regions, they may freeze out from lack of oxygen when heavy snows block sunlight penetration. Only in marshes with water-control structures can this species be controlled, and then it presents recurring challenges to management. Poisoning has been resorted to regularly for control, but the cost of poisons and energy costs for application have made this method uneconomical.

Another species of carp, the grass carp or white amur, now is widely used throughout the South for weed control. Some state agencies only allow such introductions in closed basins, but use of such biological control agents for reducing the exotic water weed hydrilla has been so successful that such efforts are likely to continue. The unique spawning of this fish in streams makes its natural reproduction in lakes unlikely, and sterile hybrids now have been developed for weed-control efforts. However, if adults of this species get into marshes, it is likely that the grass carp will compete with waterfowl and other fish because it does utilize submerged water plants and invertebrates (Gasaway and Drda 1977).

A number of species of marsh or marsh-edge plants have been introduced into new habitats from elsewhere in the world (Elton 1958, Ranwell 1967, Smith 1964). These include some serious nuisance species such as water-hyacinth, Eurasian water milfoil, and hydrilla, which have cost millions of dollars in control programs here and elsewhere in the world. Some of these introductions are results of accidents associated with the movements of humans; others were intentional but misguided acts, and will forever influence natural values of our wetlands and mandate management. We must discourage these intentional introductions.

The Limits of Management

As wetlands decrease and user demand increases, efforts increase to keep productivity of wildlife high. What is the potential of management and what are the limits? Obviously, biologists understand general concepts of plant succession, wildlife habitat selection, and niches of certain groups—but there are many unknowns. Because all environmental influences are not known and are not controllable anyway, management is not always possible—nor are outcomes predictable. Numerous observational studies have been completed, but experimental investigations with tightly regulated controls are limited. Field studies are difficult to control, and the usual pattern is to analyze a before-and-after situation. But many of the elements would change over time in a project of that kind, because years of study must be involved and year-to-year variation is common. More precise experiments are possible only when experimental units are identical, where experiments can be duplicated concurrently, and when untreated controls can be observed simultaneously. Such units now have been constructed at the Delta Waterfowl Research Station in southern Manitoba with the help of Ducks Unlimited of Canada. Additional experimental wetlands of other types have been built and are producing information that will aid in wetland establishment over a wide range of conditions (McKnight 1992). Only with such data can precise modeling and prediction become a reality.

All too often we must move from initial observation or preliminary research to an operational stage in the interest of cost saving or in response

Marsh management units were constructed at the Delta Waterfowl Research Station, Delta, Manitoba, by the station and Ducks Unlimited, Canada. These units are designed for the experimental study of marsh ecology by permitting replicated experiments of such manipulations as water level, mowing, and fertilization.

to emergencies relating to the wildlife resource. This is not good science, but benefits can accrue from these operational programs if managers view themselves as researchers as well. It is rarely possible to repeat any management operation with precisely the same timing, water conditions, and temperature, so each process is unique and requires evaluation. Such observations will lead to better data, higher accuracy, and more predictable results. Each management operation should be part of a long-term program with preliminary observations, records of the actions taken, and a follow-up measuring successes—regardless of how superficial the data may seem.

But there are other limits on managed systems: In general, we tend to expect too much from technologies (i.e., producing more wildlife on each area), and hope that they can save us from sins of the past or present, by accomplishing even such things as replacing major losses of habitat. This is an unrealistic goal for management. We can enhance production, but major increases in productivity due to management usually influence single or a small number of wetlands, not those of large wetland regions.

Management programs require (1) an assessment of what is desirable and good for wildlife; (2) some observations of natural succession and other wetland processes that drive the system, either under different conditions or with several wetland types as examples; (3) some logic in assessing important environmental influences, such as hydroperiod and water depth; (4) premanagement assessment of the situation; and (5) some modest experimentation. Armed with these concepts, we should be able to achieve fair success at managing a small marsh area to improve cover diversity and enhance wildlife populations. (See Appendix B.)

CHAPTER 9
Marshes and People

Loss of Marshes: The Good and the Bad

As people—builders, farmers, and engineers—moved across North America, the marshes and wet-prairie soils stood in the way of tillage, railroad and road construction, and other land uses. Drainage was the answer, and millions of acres of land are now underlain with tile, intersected with ditches, and built over. Estimates of the original wetland acreage in the 1780s range from 211 million to 221 million acres in the contiguous forty-eight states. The original percentage of land in wetlands in the lower forty-eight was 11%, while in Alaska and Hawaii it was 45.3% and 1.4%, respectively. Based on the National Wetlands Inventory of the U.S. Fish and Wildlife Service, estimates from the mid-1980s indicate 104 million acres of wetlands remaining, or a loss of 53% (Dahl 1990). Percentage loss by state is impressive but can be misleading because actual acreage varies so much. States losing over 50% include: California, 91%; Ohio, 90%; Iowa, 89%; Missouri, 87%; Indiana, 87%; Illinois, 85%; Kentucky, 81%; Connecticut, 74%; Maryland, 73%; Arizona, 72%; Oklahoma, 67%; New York, 60%; Tennessee, 59%; Mississippi, 59%; Pennsylvania, 56%; Idaho, 56%; Delaware, 54%; Nevada, 52%; and Texas, 52%.

Although Alaska has had only a minor loss from 170.2 million to 170 million acres, I suspect that many were the higher productivity wetlands in valleys. Hawaiian wetlands reputedly have declined from 58,800 acres to

51,800 acres for a 12% loss, but this may be highly inaccurate because of the islands' early settlement and our lack of information.

Thirteen of the forty-eight contiguous states once had 5 million or more acres of wetland and have lost over one-third of their original acreage. Listed below is the name of each state with original wetland acres in millions, followed by current acres in millions and percentage loss:

Florida 20.3	= 11.0	(−46%)
Louisiana 16.2	= 8.8	(−46%)
Texas 16.0	= 7.6	(−52%)
Minnesota 15.0	= 8.7	(−42%)
Michigan 11.2	= 5.6	(−50%)
North Carolina 11.0	= 5.7	(−49%)
Mississippi 9.9	= 4.0	(−59%)
Arkansas 9.8	= 2.8	(−72%)
Wisconsin 9.8	= 5.3	(−46%)
Illinois 8.2	= 1.3	(−85%)
Indiana 5.6	= 0.8	(−87%)
Ohio 5.0	= 0.5	(−90%)
California 5.0	= 0.4	(−91%)

Positive benefits of this drainage are the exploitation of some of the richest soils in the world for the food resources and trade products that constitute our high standard of living. Among the detrimental consequences of this drainage were losses of trumpeter swans, cranes, geese, harriers, curlews, and other wetland wildlife. This contrast is reflected in the names of the towns of Plover, Mallard, and Curlew in an area of northwest Iowa where few marshes remain and the myriads of wildfowl are gone, but where agricultural production is stable and valuable.

While smaller in total acreage, the loss of wetlands in the four major prairie pothole states is frightening in terms of potential impact on breeding habitat for waterbirds:

Minnesota 15.0	= 8.7	(−42%)
North Dakota 4.9	= 2.5	(−49%)
Iowa 4.0	= 0.4	(−89%)
South Dakota 2.7	= 1.8	(−35%)

Add to those the losses in the eastern Great Lakes wetlands of Wisconsin, Michigan, Ohio, Illinois, and Indiana, listed above, and waterfowl

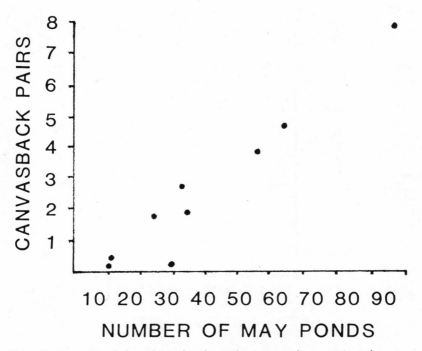

Figure 13. *Data on canvasback populations show that numbers increase in direct proportion to the numbers of breeding marshes available (after Sugden 1978).*

and other waterbird populations cannot be expected to remain the same with a 40% loss of wetlands, or a decline from 66.4 million to 26.9 million acres.

Actual population data on waterfowl from 1955 to date show major declines in most species that nest in the Prairie Pothole Region (U.S. Fish and Wildlife Service and Canadian Wildlife Service 1992), but cause and effect are assumed rather than demonstrated. Losses of the larger birds such as swans, geese, and cranes are evident throughout most of the prairie. Even more difficult to quantify are losses in population size, species composition, and species richness of other species of waterbirds for which surveys have not been made. The species composition tends to remain the same, but we recognize that the numbers of individuals must have been greatly reduced by such extensive wetland losses.

Some measure of bird loss is shown in comparing wet and dry cycles of wetland study areas with population size of ducks where good annual population estimates are available. Data for mallards and canvasbacks especially

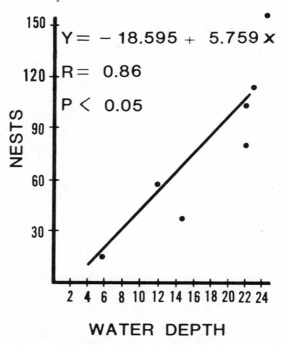

Figure 14. *Data from Dewey's Pasture Wildlife Management Area in northwest Iowa show that the increase in blue-winged teal populations was directly correlated with the increase in water levels on a single pond that was representative of water trends in the entire area (Weller 1978b).*

demonstrate that populations are highest during and following the wet years, when more wetland breeding sites are available (Fig. 13) (Sugden 1978). Fluctuations of water conditions within a given wetland or wetland complex show similar impacts on population size (Fig. 14) (Weller 1979). We can use data from such studies to predict what future populations of ducks would be if certain types or percentages of wetlands were permanently drained. In one northwest Iowa study area, for example, drought dried all small wetlands just as drainage might do, resulting in an 80% decline of blue-winged teal populations (Weller 1979). Similar data exist for the effects of drainage on muskrat populations, where declines exceeded 90% in an experimental study (see Fig. 6) (Weller and Fredrickson 1974).

Other wetland values lost include water retention, purification, and flood prevention; soil conservation through reduced wind action and erosion control along shorelines; and nutrient concentration. It has been difficult to make an economic case for preserving marshes against the values resulting from other land uses, especially when analyzed on a short-term cost-benefit basis (Anthony 1975). Social and political as well as economic influences determine whether drainage occurs. Often, these reasons are personal and shortsighted whereas values of this magnitude should be judged only on a long-term basis, using estimates of net benefits to all users, direct and indirect.

Evaluating Marshes

Marshes and other wildlife habitats have long been viewed almost entirely for their esthetic values, and biologists have resisted economic analyses even when fighting to preserve areas from destruction. However, economic considerations have legislative weight, and efforts were made first to value components of the marsh, such as ducks or fish, because it is easier to value a "product" than it is to value esthetic and recreational aspects. Economists seem to use two approaches for evaluating natural resources such as wetlands; either they try to develop novel approaches to placing dollar values on them, or they consider them as entities on which values cannot be placed directly, but about which decisions are made by society.

Wildlife recreational value has been measured in terms of expenditures for sport hunting, fishing, and watching wildlife. The U.S. Fish and Wildlife Service periodically assesses these values to the U.S. economy and suggests that hunting and fishing are billion-dollar industries (U.S. Fish and Wildlife Service and U.S. Department of Commerce 1993). Other economic studies have tried to assess the value of a local area to the economy. Horicon Marsh, in central Wisconsin, is a large marsh where drainage proved unsuccessful. It was restored by state and federal efforts, and its goose management program was so successful that game harvest and goose watching became major local activities. Hunters bought hunting rights from farmers in the area; even more people came to see the geese and spent considerable money in the area. In 1960–1961, it was estimated that communities near Horicon Marsh gained a half-million dollars in sales annually from marsh-associated recreation (Keith 1964).

Another economic valuation involves such things as "willingness to pay," a technique that some economists feel can be used to assess esthetic experiences and social values as well as impacts on communities. One such measuring device is a questionnaire that postulates bidding on items or issues; the respondent indicates the limits to which he or she will go to participate in an experience, such as seeing a swan or whooping crane, enjoying the sight and sounds of masses of migratory waterfowl, or watching a beaver building a dam at sunset. Another system involves a comparison of travel costs by different users at different distances, with the differential providing an estimate of how increased costs of travel will decrease participation or use (Bart et al. 1979). Unfortunately, few of these methods seem to have been used on marsh valuation except for evaluating waterfowl hunting or some other consumptive use. An excellent study of wetland value and duck value in relation to duck populations attempted to model mallard harvest in relation to wetlands needed for the population (Hammack and Brown 1974). Studies specific to marshes have included various visual, cultural, and economic valuations (Larson 1971, 1975), basing economic estimates on the most valuable alternate use of the resource (Gupta and Foster 1975). Some authors have explored the use of assessments that include the areas' potentials for sewage waste treatment and total "life support," such as the production of oxygen, carbon dioxide, and carbohydrates (Gosselink et al. 1973). Other economists find these figures unrealistic and encourage the development of comparative figures that are more likely to be accepted (Foster 1979).

Public education through the activities of national environmental groups, local nature centers, and state and federal agencies has resulted in much greater appreciation of wetland values. In a study of public attitudes toward wildlife and wildlife habitat, 57% of the respondents disagreed with the view that we should fill and build over marshes as long as endangered species were not involved (Kellert 1979). In Florida, almost 72% of 250 persons sampled recognized the values of wetlands for wildlife, and 60% understood that wetlands control storm surges and floods. Only 20% thought they were more of a nuisance than a value.

In recent years, social action through environmental groups and government lawmaking bodies has resulted in some improvement for preserving wetlands, and perhaps more of this has occurred at state and local than at federal levels. In many cases, however, the action is to preserve marshes and other wetlands through long-range planning, avoidance, or mitigation for losses. To accomplish this type of preventive action or compensation for damages, evaluations of the marshes must be used as a foundation. The major system now in use to accomplish this evaluation is the Habitat Evaluation Procedure (HEP) employed by the U.S. Fish and Wildlife

Table 2. Functions and values of wetlands recognized in the Wetland Evaluation Technique currently used by the U.S. Army Corps of Engineers and other agencies (Adamus et al. 1987, 1991)

Functions (which also may be considered values to some)
    Groundwater recharge
    Groundwater discharge
    Floodwater alteration
    Sediment stabilization
    Sediment/Toxicant retention
    Nutrient removal/transformation
    Production export
    Aquatic diversity/abundance
    Wildlife diversity/abundance

Values (which do not perform functions within the wetland)
    Recreation
    Uniqueness/Heritage values

Service. By using various models that assess the suitability of the habitat (formalized as a Habitat Suitability Index) for a certain species or group of species, a per-acre index is established that is used in mitigation actions where replacement must simulate the total quality and not just the acreage. For example, twice as much land would be required for compensation if its index value was only half that of the land to be lost.

But HEP focuses almost entirely on wildlife values, and the most recent effort to provide rapid, qualitative evaluation systems for wetlands was developed for the Federal Highway Administration and subsequently modified for the U.S. Army Corps of Engineers as the Wetland Evaluation Technique (WET) (Adamus et al. 1987). This system takes into consideration the various natural and valuable functions of the wetland as it relates to society, such as water-quality enhancement, flood control, primary productivity, as well as food resources and habitats for fish and wildlife, and passive recreation, fishing, or hunting (Table 2). This system involves a simple quantitative scale: A wetland can be judged on multiple functions and values, or two marshes can be compared and the unique qualities of a wetland will stand out when compared with any other area. In this way, we do not ignore these hard-to-measure but valuable properties of the ecosystem that will be costly in the long term to lose. We also can place other human values on them as well, for it is clear that a shoreline property on a lake or ocean brings a higher price than on a wetland, but both are far above other natural communities in price, because of their rarity.

Regulators generally do not favor an approach involving measurement and replacement of functional value because it is difficult to accomplish quantitatively and realistically in the time frames allowed, and developers do not like it because it makes dollar comparisons less relevant. Regardless of the parameter used to measure the value, the functions of a wetland are likely to continue for thousands of years in some form. The functions of a new factory or house are relatively short-lived, but they may have important local economic impact, and this tends to override long-term nonmonetary values. The people must judge what they deem most important, and it will vary it different places at different times. Nevertheless, we cannot make sound decisions on long-term issues of great societal importance unless we understand what is being lost and why.

A recent suggestion favored by some politicians and consequently being evaluated by federal agencies is designed to aid in making decisions quickly on use of wetlands and wetland sites. It proposes to "categorize" wetlands by qualitative standards such as high, medium, and low, for example. Unfortunately, some may view the system as one intended not to protect the best, but to facilitate modification or destruction of the lower categories of wetlands with less regulatory effort. This might well speed development but also would result in loss of the "drier" wetlands shown to be very valuable for many functions. But who will judge, and on what basis? Surely, we must understand how wetlands function in the natural operation and self-perpetuation of our local landscape before we eliminate more than the remaining 46% of the U.S. wetlands. Can this modest bit of a landscape type—less than 5% of the geographic area of the contiguous forty-eight states—really be that precious for alternate uses that we have to use every acre?

Negative Aspects of Marshes

Although marshes have many, often subtle, values, no one will deny that marshes can create problems for various reasons: excessive water where it is not wanted, unsuitable soils for farming (Anthony 1975) or construction, flocks of granivorous birds such as blackbirds or ducks, and, possibly, "weed banks."

Even when tilling and wetlands seem compatible, there are those years when the snow or rain has been heavy, and marsh water levels rise to flood out farms, buildings, and roadbanks. Little can be done in such situations, as tiles and ditches are full, pumps have no place to dump the water, and cutting of impediments may do more harm than good. Water-logged soils are no good for construction, and suitable foundations are difficult to build under flood conditions. Road engineers never seem to tire of meeting the challenge, but high water and muskrats usually win out, and many old roadways near marshes have been abandoned.

The mosquitoes at marsh edges can be a serious nuisance as well as be carriers of disease. However, mosquitoes could thrive without marshes, because of the many species that occur in diverse habitats. Seasonal flood-waters are especially significant as causes of mosquito production. Drainage has not proved to be an effective means of eliminating production sites because many species breed in minimal amounts of water, but some adverse effects on wildlife have resulted from such efforts (Stearns et al. 1940). Control programs generally are restricted to urban areas where newer, short-lived insecticides are used.

Generally a minor problem in rural areas, marsh odors can be a material issue in urban areas. Because marshes are traps for organic matter, and much decomposition occurs in late summer when water levels decline, rotting vegetation and sulfur-laden gases are not well regarded.

Among the greatest challenges to wildlife managers have been the difficulties inherent in farming adjacent to wetlands, where blackbirds and ducks concentrate. Especially vulnerable crops are wheat and barley in drier prairie regions, corn and sunflowers in the north-central states, corn in the Southeast, domestic white rice in California, Arkansas, Louisiana, and Texas, and cultivated wild rice in Minnesota, Wisconsin, and Manitoba. The red-winged blackbird is perhaps the most numerous bird in North America and breeds in a variety of habitats. Redwings flock in late summer and early fall and feed on ripening corn, sunflowers (Stone and Mott 1973), and rice (Meanley 1971) in the "milk" and seed stages. In wheat, barley, and oats areas where the grain is swathed to dry, as is still common where it does not ripen evenly, blackbirds and ducks both gather to feed. A combination of heavy rain and downed grain is an open invitation to trouble: Much grain is lost from the heads to ducks "puddling" rather than eating. This depredation is so serious that complete crops have been eliminated in a few rainy days of late July or early August in the southern parts of the Prairie Provinces of Canada. Various techniques have been used to deter flocks, such as acetylene exploders or gunners to frighten birds away, bait sacrificed to detract birds from more important crops,

An ephemeral pothole was plowed in the spring but later reflooded. Such mud flats can be favorite places for shore-birds, but reduction of organic matter may reduce their productivity in another season.

A flooded wetland shows (a) edges attractive to loafing dabbling ducks and (b) remains of an old road that formerly ran through the marsh and was abandoned probably because of the maintenance problems resulting from high water and muskrat burrows.

or chemicals that repel blackbirds from crops (e.g., Methiocarb), but the losses go on and the expenses of farming increase. Because losses often are a matter of timing of ripening, harvest, rainfall, and bird flocking, they usually do not occur every year. Some areas have used crop insurance to minimize the impact on a certain farm in a single season. Nevertheless, the problem is real and, in some areas, may force people to limit bird population levels. This is another example of conflict between humans and nature where humans always expect to win—and do not.

Water-Level Modification

In some cases, marshes are lost because of misdirected conservation or management efforts. Increased water levels often are cited as desirable goals because the general feeling is that such marshes are more attractive to wildlife. Certainly, they are more attractive to humans and fish. High water levels reduce vegetation after several years and, depending on the starting point, tend to create lakes, not marshes, and aquatic rather than semiaquatic communities. An example may help clarify the perspective and the problem.

Throughout the glaciated Prairie Pothole Region, lakes occur that locally are called marshes, and marshes that are called lakes. Such was the case with Little Wall Lake in central Iowa. There are many "wall" lakes in the Midwest because they were formed when their drainage was "walled" off by a glacial moraine, or dammed by ice action against the shoreline later. But they vary from marsh to lake condition, owing to the dynamic water cycles discussed earlier, and their name depended on their condition at the time of survey and naming. This situation may produce local public concern over the lake that looks like a marsh, and many such wetlands have suffered from public pressure to make sure that the body of water looks like what its name is! Such was the fate of Little Wall Lake.

In spite of its sizable acreage (230 to 275 acres), Little Wall Lake was dry in about 1894, 1904, and nearly so in the 1930s and the mid-1950s. But when full, it was a diverse and exciting marsh community, with some sizable water openings. An Iowa Geological Survey report (MacBride 1909) dramatizes both the human conflict and the character and richness of the "lake." "Had it depth, Little Wall Lake would be the attraction of the landscape, but its shallowness makes it simply a great marsh filled from side to side with aquatic plants. The margins are dark with sedges. In the middle, the cattail lifts its blades undisturbed, while over the deeper waters the pond lilies spread their broad leaves like inverted shields and star the surface with flowers. Innumerable birds fill the air with strident, unmusical sounds; ducks steer their miniature fleets about; mud hens wade among the Calamus roots; blackbirds cry as if life depended upon unceasing verse; the tern hovers above the more open waters or sits upon the sand as if by the sea; the bittern sits among the reeds, bill straight up, more like an inverted stake than any "stake driver'; and over all, in the evenings, clouds of insects-mosquitoes make gray the air on every side."

The "mosquitoes" were, of course, the midges that swarm on warm and quiet nights in all marshes, and they and the birds described still remained (at lower population levels) when Paul L. Errington, authority on muskrats and marshes, first showed me Little Wall Lake in 1957. It was suffering from several years of drought, so that the marsh vegetation was almost dry. But there was a large area of open water, the consequence of dredging the area in 1953. This was an area of about a hundred acres that attracted numerous motorboats.

Our observations of the bird life and habitat dynamics of Little Wall Lake spanned nearly eight years (Weller and Spatcher 1965), and in their last year were in synchrony with a project that raised the lake level and inundated all vegetation (Fig. 15). The effects of creating a full lake from this former half-marsh caused the loss of most muskrats, mink, and bird life. A few diving ducks and coots still stop in migration, but it is not a breeding place now. Recreational activities flourished: Fishing was great, as is usual with newly flooded vegetation; a camping area was developed and was constantly filled; and the motorboats and sailboats increased the head count of users, with significant local economic benefits.

It could have been worse. Many people had recommended drainage, a pattern in this area, which contains some of the deepest and most fertile topsoils in the world. But some local residents had favored keeping the water, and in 1917, it had been proposed that the "wall" be raised to create a dam, that the water level be raised, and that forty acres be dredged. Fifty years later, it all came to pass. Little Wall Lake became a lake at last!

A few mourned the passing of the marsh, but most people were delighted to see water, if only to park by it momentarily, and viewed the whole project in a positive light. Most saw this project as a conservation program, and it was financed in part by funds dedicated for that purpose at both state and county levels.

The same effect is felt in areas where small marshes are drained into large ones. Wildlife habitat is lost in both places. Small marshes become farm-

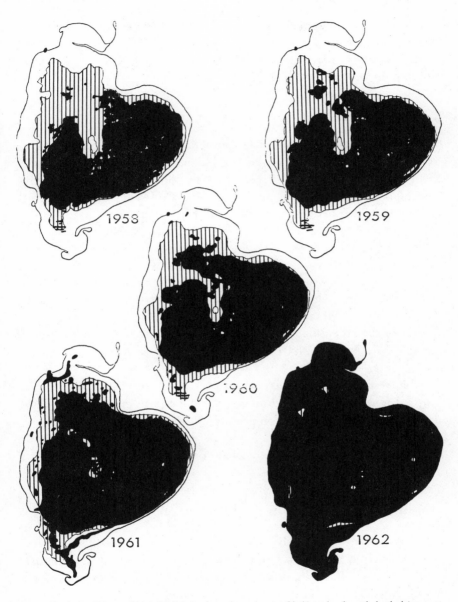

Figure 15. *A water history of Little Wall Lake shows the open water (black) produced mostly by dredging (1958), the dry emergent vegetation (pattern), and the sedge or upland edges (white) (Weller and Spatcher 1965). A gradual flooding improved cover-to-water interspersion in 1959 to 1961 and was reflected in increased bird numbers and species richness. Finally, artificially elevated water levels in 1962 eliminated most nesting birds except for a few red-winged blackbirds and grackles nesting in small trees along the perimeter.*

land, and large marshes become lakes—good for fish but not for waterbirds or muskrats. The effect on wildlife is not as great as when all areas are drained, but the entire character of the area is changed, with a concomitant reduction in wildlife production and species richness.

Water-Level Stability

Through a misunderstanding of recommendations directed at preventing drying or flooding of waterfowl nests (Hochbaum 1944) or muskrat lodges (Errington 1937) during the breeding cycle, some marsh managers strive for stability of water levels. Observations of nest losses suggest that such stability does induce better nest success and may be a suitable management goal during the nesting season—but not during the whole year. Seasonal and year-to-year variations in water levels are normal events resulting from variations in rainfall, and most marsh organisms are adjusted to these changes. Somehow, we seem to expect maximal wildlife and plant production every year, whereas evolutionary adaptations are a product of optimums or averages; there is security in the flexibility of not being at one extreme or the other. Stability of vegetation is the product of an average of year-to-year conditions. In some years, plants gain in density or coverage; in some years, they lose.

Stable water levels throughout the year have several negative effects. First, nutrients seem to be tied up in organic debris in the substrate, where anaerobic conditions slow decomposition. Reduced emergents result in reduced nutrient pumping into the system, although some submergents do this quite well. When the supply of nutrients is reduced, production of other aquatic organisms declines. Second, plant communities that require low water levels for either seeding or survival suffer population losses that bring about lakelike conditions. In such cases, marsh life is replaced by lake life.

Marsh managers do not always have the option of regulating water levels, because control structures or water supply do not permit it. In the arid West, where wildlife managers compete for water with other users, water may be available only early in the year. If it is held at high levels for retention into the fall for hunting or other reasons, the effects on vegetation often are disastrous because seasonal and year-to-year variation is essential for holding diverse types of vegetation and making available nutrients for the growth of plants and invertebrates.

A large seasonal wetland (top) was drained early in 1979 and planted to forage grasses. This was one of several wetlands that drained into a still larger semipermanent marsh (bottom) that then was flooded beyond its usual banks.

Marshes for Water Purification and Energy

There is well-documented evidence that water that flows through a marsh comes out cleaner and less enriched with most nutrients than when it entered (Kadlec 1979). Engineers are taking advantage of this cleansing principle, using both natural and cultivated stands of emergent plants, such as cattails, for settling basins and nutrient traps. Some other experiments involve using cypress swamps (Ewel and Odum 1984), northern bogs, and coastal tidal salt marshes for purifying human wastes from small communities. In natural systems, there is strong indication that species diversity of plants is reduced by sewage or other pollutants, and that floating plants may be enhanced, as is true in late stages of marsh deterioration when nutrients are more abundant. We certainly do not know and can not adequately predict the long-term effect of such concentrated nutrients on energy flow, species diversity, successional trends, and water quality of a marsh. In agricultural areas, runoff includes significant amounts of pollutants such as fertilizers, which could induce eutrophication of wetlands (Neely and Baker 1989); herbicides, which will influence plant species composition; and insecticides, which may modify invertebrate populations.

It is obvious that the marsh is a forgiving and adaptable system, but it should not be viewed as a dump for excess nutrients, any more than it should be used for old cars and other human debris. The safest approach is to avoid use of natural wetlands as purifiers or to regulate the nature and extent of the effluent. However, it is well known that a variety of aquatic plants can be used in artificial stands managed specifically for processing human and livestock sewage as well as other animal wastes (Hammer 1989, 1992).

Aquatic plants also have been used experimentally for nutrient removal of waste water, with subsequent conversion of this nutrient to plant biomass that can be used for livestock food or for fuel. Some plants, such as duckweed and the introduced submergent hydrilla, produced fifteen and seventeen U.S. tons (13.6 and 15.4 metric tons) of biomass per 2.5 acres (hectare) per year in subtropical Florida, which is comparable to production of temperate terrestrial crops. Water-hyacinth had the highest production recorded for any crop, ninety-seven U.S. tons (88.0 metric tons) per 2.5 acres (hectare) per year. Based on this production, it has been estimated that a 1,000-acre (405-hectare) water-hyacinth farm in the southern United States could produce 1,012 BTU of energy as methane, and remove nitrogen and other nutrients from waste water for a population of 700,000 people (Ryther et al. 1979).

Reed, widely used as a building material in Europe and Iraq, has been used in Sweden for burning as a substitute for fossil fuels. With proper selection of cutting area and patterns, reed harvest could be a marsh management tool. Cattail use for energy has been promoted in Minnesota where other natural energy sources are scarce (Andrews and Pratt 1978).

Hence, several possible human uses of hydrophytes have been explored and have potential for the future. Some may adversely affect the marshes; others may not. First and foremost, we must protect the marsh system that creates these plant riches for the wildlife dependent on them.

Wildlife Users: The Watchers

According to most surveys, wildlife watchers are growing in numbers faster than wildlife users. Obviously, anyone can be a watcher, and current environmental concerns have heightened the public awareness; national programs have enlightened people's attitudes, and nature study centers and television programs have excited youthful minds. Like the people of several European countries, we have become more knowledgeable about and more interested in wildlife. The reduced costs of binoculars, telescopes, recorders, and telephoto camera equipment have given many people the chance to be serious students of wildlife, and wetlands often are a focal point for such study because wildlife concentrates there. Bird watching is even becoming socially acceptable.

One of the best ways to enhance the appreciation of wetlands is to get people into them. At nature study centers, some people are donning hip boots and going into wetlands with plankton nets, magnifying glasses, and cameras. But for most people, such efforts are unlikely to be undertaken, so alternate systems are used. One of the most helpful is the construction of a walkway through a wetland so that watchers are brought nearer to the sights, sounds, and actions of the animals, by moving through the vegetation. Although doubtless some damage does occur during construction of such boardwalks, the price is worth the result for a selected few marshes.

Wildlife Users: The Takers

Hunters, trappers, and, sometimes, fisherfolk are the major users of marsh wildlife. Hunting of waterfowl, rails, and snipe (in some states) is a principal sport, and reduced numbers of wetlands have created excessive pressure on remaining areas. The pressure such "users" exerted for the preservation of wetlands preceded any action by wildlife "watchers," and the influence of these users is no less important today. Biologically, such users do little damage to either marshes or their wildlife; their conservation interests do nothing but good.

The philosophy of harvest versus nonharvest is beyond the scope of this book, but the biological issue should be at least presented. Most harvest (hunting, trapping, fishing) utilizes the annual, seasonal surplus of young that normally would be lost to predation, disease, competition, or climatic stress. Such mortality tends to be directly density dependent; that is, mortality is greatest when the population is high, lowest when the population is low. For many species, human harvest replaces other causes of mortality, and does not reduce the population below the normal spring breeding level (Rogers et al. 1979). However, maximal harvest must be regulated so that populations have a surplus for protection against disasters such as storms that affect animals regardless of their density.

Because certain species of wildlife are almost entirely associated with wetlands, the major national harvests of such species as ducks and rails can be viewed as products of wetlands. Muskrats, too, have their maximal densities in marshes. A summary of such harvest suggests that about 10 million ducks are taken by hunters annually in the United States (Chabreck 1979), but no monetary value is generally assigned them as they are not sold. Recent summaries indicated a kill of about 11 million ducks and 1.7 million geese taken in 1987 (Martin et al. 1989). Furbearers, however, constituted a $35-million harvest in 1974–1975, with an industry value that would be far greater. Fifty percent of the muskrat harvest occurs in the Midwest, with the largest numbers in Wisconsin, Minnesota, Ohio, and Iowa. Almost all nutria (97%) were taken in Louisiana, and 58% of mink were taken in the same states as were muskrats (Chabreck 1979).

Endangered Species and Endangered Habitats

The problems of losses of endangered species have stirred the public interest as have few other wildlife issues. The approaches used to resolve population problems of such species have been unique and often effective. But many endangered or threatened species are rare mainly because their habitat has suffered severe losses. There are numerous wetland species among the endangered forms, including some conspicuous species that have been seriously impacted by human activities: snail kites that reside in the Florida

Everglades; whooping cranes that nest in the Northwest Territories of Canada and winter in coastal Texas, but which once nested in marshes well south into the Midwest; and trumpeter swans that are residents in Alaska and British Columbia, but also once occurred in midwestern marshes. Several species of rails, especially the Yuma clapper rail in the Southwest, are rare or endangered, but the cause is uncertain. The Houston toad is a small wetland species, now considered highly endangered, that is being crowded out by drainage and housing development.

Clearly, the way to keep more species off the endangered species list is to preserve habitats. When habitats become endangered, so will many members of the community, plant or animal. It is difficult to assess whether whooping cranes and trumpeter swans were ever abundant breeding birds in the Midwest and what caused their disappearance, but both require large ranges for feeding and rearing young. Wetlands in the region now are mostly relicts, often of small size, that are able to hold those adaptable species that have small ranges and diverse food choices.

The Contribution of Wetlands to Environmental Science

One cannot evaluate the preservation of natural features for scientific study without relating it to human needs. We constantly search for our role in nature as well as in human society, and this search for understanding includes the need to know how natural ecological communities function. As scientific approaches, philosophies, and techniques change, we can dig more deeply into such systems or interpret them more fully. It is vital that representative natural systems be preserved for study as well as for their other values. Moreover, these cannot be tiny, isolated, and atypical units; some components of the marsh community (e.g., larger birds) are not attracted to tiny units. In addition, the rarer plants may well be lost unless large, diverse communities are preserved.

Marshes are especially suitable ecological units for observation of natural phenomena by students, and ideal for the scientific study of systems. Because wildlife often concentrates in such energy-rich units, study of these species has unexpectedly enhanced our understanding of sociobiology, mating systems, cyclic phenomena, habitat relationships, population regulation, and other biological facts. Scientists have only scratched the surface; much research remains to be done on the dynamics of the plant community in wetlands, in studying nutrient flow, and in developing data bases, allowing us to interpret and predict events in the ever-changing freshwater marsh.

CHAPTER 10
Marshes for the Future

The Human Need for Water

For such terrestrial creatures, humans are perennial seekers of water. Coasts, rivers, and lakes are rimmed with human dwellings, boat landings, overviews, and bathing or wading areas. If bodies of water are absent, we build them. Much of this search is, of course, for the essential compound of H_2O itself, but there seems also to be an esthetic and visual need for the openness, motion, and reflection of a body of water. In general, the visual clue is essential; just knowing that water is there is not satisfactory. Thus, a marsh has less "water value" for most people than does a lake. Marshes have not featured as strongly in social or esthetic importance as other natural features have. Except for the Everglades, no wetland has been the focus of a national park although many parks do have exceptional wetlands; wetlands have been almost ignored as wilderness areas (Fritzell 1979). Until recently, few would consider building a house overlooking a marsh, whereas lakeshore or riverside property brings the top price.

This attitude does seem to be changing, however. In part because of the wildlife commonly found around marshes, in part because of the quest for open space, urban and rural residents seem now to have greater respect for wetlands, and some even seek them out. Wetlands add diversity of color, pattern, and form to rather uniform urban areas or fields; they attract a diversity of wildlife that varies with the seasons. The recent outdoor trend has brought canoers in summer and cross-country skiers and snowshoers in winter to marshes where only hunters and trappers passed before. Many still encounter the track of mink, weasel, or fox without knowledge or recognition, but we are learning to seek the excitement of the usually unseen. In areas where these wetlands are common, public education has increased, and public awareness and concern for the wetlands and their wildlife are growing.

Wetlands in Conflict

In most areas, our desire for water takes precedence over our wish to understand wetlands or their fauna. Where semiopen marshes occur, we dredge and deepen, or where less water is present, and marshes cannot be filled or drained (for legal reasons, if no other), we build dams to stabilize and deepen the water. This increases property value or makes water areas more attractive to boaters, fisherfolk, campers, and picnickers. Those who have read to this point now know that such "stability" will result in reduced diversity and production of wildlife and, eventually, even fish. But public pressures and involvements being what they are, many areas often are changed at the hands of humans, usually to the detriment of the flora and fauna.

For many years, government policy on wetlands in rural areas was ambivalent: The U.S. Soil Conservation Service in the Department of Agriculture encouraged, engineered, and even subsidized drainage of wetlands in the interest of neater, more efficient, and more productive farming (Leitch and Danielson 1979). The U.S. Fish and Wildlife Service, in the Department of the Interior, has encouraged the preservation of wetlands, and leased and purchased them for waterfowl and other wildlife. The U.S. Army Corps of Engineers, U.S. Bureau of Reclamation, and U.S. Bureau of Land Management developed flood control and irrigation programs that eliminated wetlands. Such conflict now is less of a problem because of the national policies of these agencies, culminating in 1977 with a presidential executive order to minimize wetland losses or degradation, and with congressional actions to encourage preservation of wetlands for alternate values. The Department of Agriculture has had a special Water Bank Program that included wetlands, and stream channelization that seriously impacted wetlands (Choate 1972, Erickson et al. 1979) has been reduced. The Soil Conservation Service now is involved in wetland delineation and even provides guidance on restoration and creation of wetlands (U.S. Soil Conservation Service 1992). Moreover, drainage of agricultural and forest lands that once were excluded from federal regulations (Harmon 1979) is discouraged and may result in loss of subsidies under the "swampbuster" provision of the Farm Bill. The Corps of Engineers now bears the major responsibility for protection of wetlands from abuses such as dredging and filling and issues permits under Section 404 of the Clean Water Act.

However, there are many problems, and policies are inconsistent from region to region and from state to state. There is still too little concern in some agencies for smaller and thus drier wetlands, which are especially vulnerable because they come and go with wet and dry seasons. During the dry years, installation of drainage pipes is easy, and losses occur indirectly even where marshes are legally protected. Unfortunately, a few major wetland areas are still threatened by government programs for irrigation or flood control.

Some states and metropolitan areas retain ownership of wetlands and also have laws preventing filling or drainage of most larger wetlands. Many do not, however, and local losses continue. In addition, siltation due to poor management of upland cover can fill wetlands to a point where they lose their identity, but no legal action is possible. Local interest and attitudes are especially important, because of both the potential control states have over wetlands and the support they can give to federal rulings and policies. Over half the states now have enacted legislation to protect wetlands (Osvald and Belin 1979, Rosenbaum 1979), several have no-net loss laws that require creation if wetlands are destroyed, and several states have tax incentives to landowners to induce preservation. In other states, county boards or the governor have prevented the sale of wetlands to federal agencies wishing to preserve in perpetuity (Kusler 1983).

Where modifications of wetlands involve a water control structure, excavating, or filling, the Corps of Engineers must provide a permit under Section 404 of the Clean Water Act (Horwitz 1978, Osvald and Belin 1979). The Environmental Protection Agency has veto power over such permits and seems to be using that authority more regularly.

At both national and state levels, the tendency is to delegate control over wetlands to localized boards involved in land-use planning and zoning (Larson 1975). This policy may have good and bad aspects, depending on local interests and attitudes. Whatever the level of authority, developers and conservationists are pitted against one another in each move to modify or drain a wetland, and an environmental impact statement (EIS) often is required. Each EIS tends to be independent of others. In most cases, guidelines are obscure, cause-effect relationships are uncertain, and studies "spawn" reams of paper, while the basic issues are rarely uncovered. Some good comes from such studies, but "paper science" is not the way to discover the values of a wetland or of any other natural resource. We calculate benefit–cost ratios for natural treasures that are priceless and irreplaceable; we compensate for losses by mitigation that may consider only the size of areas, when they are of different types of habitat and serve different purposes. Moreover, many of the areas created cannot be evaluated later because of personnel shortages, so the mitigation may not result in a functional and permanent replacement. The concept of "mitigation banking" may have some positive benefits in providing better long-range planning and commitment for especially valuable wetlands (Zagata 1985), but it also presents some challenges to our evaluation and enhancement techniques. Unfortunately, each case tends to be judged on its own merits, when in fact the cumulative value of one wetland as it is related to others locally or regionally may be the most vital and unappreciated issue.

And who should decide? If raising the level of a marsh floods an adjacent landowner or county, local legal action is possible. If it reduces the wildlife diversity or otherwise influences biological characteristics of the area, what effective actions can be taken? In legal cases, county action may override city action, or state law dominate county rulings. But rarely does such decision have an adequate scientific basis, and each court calls its own set of authorities to meet the questions and challenges, attempting to resolve issues that should involve major policies as well as localized data input. Clearly, this conflict is not easily resolved, but it can be eased by developing a national policy that would incorporate the best elements of resource conservation for the good of the nation.

Conservation Goals and Policies

One of the greatest causes of conflict in conservation programs is identifying, establishing, and communicating the objectives of the program. Thus, establishing sound practices within the responsible organization (governmental or private) is difficult, and the public is confused and concerned about the techniques used to achieve these goals.

Decision making about wetland preservation is in need of serious revision, as is decision making about other resources and our uses of them. Additional socioeconomic and biological data will enable further quantitative assessments, modeling, and predictions—and thus a better basis for decisions. But not all policy need be quantitatively, and especially economically, based. Why should not governmental agencies have policies that reflect esthetic, philosophic, and other noneconomic or nonpolitical considerations? Agency surveys have demonstrated such public value. Moreover, such agencies should exercise leadership, making use of the wisdom, motivation, and excitement of the experts they employ. National concerns and public expressions from knowledgeable environmentalists are perhaps more influential in the evolution of policy than are the efforts of government professionals.

Both national leadership and national interest in wetland conservation are more in evidence, and significant effort has been made toward a national wetland policy (Kusler 1983, National Wetlands Policy Forum 1988). Interest in wetland science has exploded as shown by the growth of the Society of Wetland Scientists and the Association of State Wetland Managers, technical conferences of interested persons at national, state, and local levels. The development of regional research centers in agencies and universities and the availability of advanced coursework are additional products of national interest in this area.

Our future land ethic must give greater priority to our natural resources such as wetlands. Current attitudes favoring the preservation of endangered species reflect public agreement with this philosophy, and it is a logical transition to the preservation of whole communities and ecosystems —not as relicts, but as functional parts of our landscape. This undertaking will test the limits of our dedication! Are such attitudes a matter of convenience or economics, and do they change generation by generation? Predictions of human population growth and its impacts make it obvious that natural resources will continue to dwindle at alarming rates (Barney 1980), unless changes evolve in our resource and population policies.

Epilogue

You, the reader, must now face a decision vital to the future of wetlands and their wildlife in North America. I hope that understanding and concern have been enhanced by this reading, and that action is the next step. This action may take many forms. For some, it may be the investment of time or money in private acquisition programs, such as those of the Nature Conservancy, or legal defense programs of the National Audubon Society, National Wildlife Federation, Sierra Club, or Environmental Defense Fund. Involvements with the North American Waterfowl Plan through some state or federal agency or through Ducks Unlimited will appeal to many hunters. For others, it may be political action at local, state, or national level. Still others may express and defend an opinion over the dinner table, an action even more effective if overheard by our children and friends. But all these acts, subtle to dramatic, must one day lead us to a national policy that will bring about the retention, management, and widespread use of one of our most unappreciated natural resources—freshwater marshes.

APPENDIX A
Some Elementary Marsh Study Techniques

Although a few textbooks and laboratory manuals describe techniques for the study of freshwater organisms (Ward and Whipple 1959) and animal populations or plant communities (Giles 1969), or show how to cover-map habitat, few of these books emphasize marshes. Examples of a few techniques for study of marshes may reduce the fear of some experimentation and observation by laypersons. In addition, there is no better way to learn about marshes than to get into them. The techniques outlined here provide an excuse to go into a marsh and examine in some detail what is there. If one is careful about pathways through vegetation and near nests, little damage is done in examining a marsh system, and much can be gained by students of any age or profession. Moreover, familiarity with such techniques aids in interpretation of environmental impact statements and other documents to which the lay public is now exposed.

Your First Steps in the Marsh

First, one needs to conquer the fear of a marsh as a bog or quicksand, where a person will sink to dangerous depths. Marshes do sometimes have muck bottoms and are soft and sticky. Marsh walking often is hard work, but even though you may fall in a time or two, you will develop "marsh legs." The only dangerous places I have encountered in a marsh have been areas of sedimentation in backwaters resulting from channelization. In fact, many marshes have very solid bottoms made up of sand or root stalks. A canoe paddle or walking stick is an essential aid.

Second, wear rubber boots or waders, or at least protective shoes and pants.

Old plant stalks can be sharp and painful, and bottles and other debris tossed into marshes in urban areas can be dangerous. Moreover, even experienced marsh professionals dislike the leeches that are abundant at some marsh phases, or the "swimmer's itch" that can be caused by a harmless but irritating larval parasite of waterbirds. In southerly areas, one must be wary of poisonous snakes.

Canoeing or boating in a marsh is hard work, but may be necessary when three feet of water and a muddy bottom make walking difficult or impossible for amateurs. Poling a canoe or boat is a technique soon perfected by those who must push through dense vegetation. Especially in the Everglades and in coastal marshes, propeller-driven airboats of various designs are used by professionals, hunters, or fisherfolk who must travel long distances in marsh vegetation or shallow waters where motorboats will not function well.

An easy way to gain experience in a northern marsh is to examine it during the winter after freeze-up. Wearing slip-resistant boots, snowshoes, or skis, one can "walk on water," gaining a most valuable perspective. Moreover, there are surprises in the activities of mink, weasels, muskrats, deer, and winter birds.

Some Scientific Philosophy

Regardless of your reasons for gathering information about a marsh, learn to do it quantitatively. It is best to use an accepted, published standard, as may be found in one of several technique manuals, but when developing a new technique, use it throughout a series of observations. If you are comparing the number of snails in two habitats, use a sampling net of the same area and mesh size, and the same number of dips or sweeps in all sampling sites so that you have comparable data. Counting blackbirds, muskrat houses, or anything similar should follow a uniform system even if there are no accepted standards.

It is also advantageous to separate observational versus experimental procedures. One can make observations, even quantitative ones, but not know whether these are typical, unless repeated observations are made. It is even more difficult to assess the causes (as opposed to effects or occurrences) of biological events without being able to experimentally reproduce them and even to replicate the experiment. Hence, experimental studies, obviously the most difficult and costly, are also most likely to yield undeniable results and to facilitate identification of causes of observed effects. Although you may not wish to involve yourself in such sophisticated studies, this scientific logic helps you to evaluate the work of others, and avoid accepting statements when they lack supporting quantitative and experimental data.

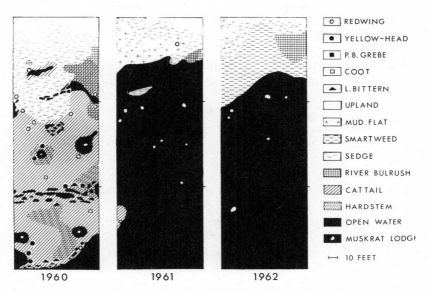

Figure 16. *A belt transect through the edge of a marsh forms a general map of vegetation and open water on which muskrat lodges and bird nests are plotted (Weller and Spatcher 1965).*

Plant Presence, Cover, and Density

Many methods of plant assessment can be followed, but the selection of a technique depends on whether one is simply assessing presence or absence of a species, or whether quantitative data are needed on the distribution and population size. Although it is difficult to obtain agreement about techniques, standardization in a single series of observations will help resolve many of the difficulties typically encountered in plant survey. Some good techniques of assessing both plants and soils are available in the various generations of the wetland delineation manual (Federal Interagency Committee for Wetland Delineation 1989). Water requirements of hydrophytes have been summarized for wetlands delineation purposes (Reed 1988), and updated information on classification for wetlands soils also is available (U.S. Soil Conservation Service 1991).

Cover mapping will provide patterns of distribution of emergent plants by species or life form. Such cover mapping is difficult if the water is deep, and in northern areas, I have found it easiest to do on ice in winter. Although some changes occur with new vegetation in the spring, this is the vegetation that greets birds on their arrival and that influences where their activities will take place. Such mapping also can be done from aerial photos if they are available in local offices of the Soil Conservation Service or associated agencies. Randomly selected belt transects or quadrats can be used when marshes are too large for total assessment (Fig. 16) (Weller and Spatcher 1965).

Detailed vegetation data usually are obtained by one or several of the following methods. First is point-count or point-intercept transect. The easiest system is to stretch a tape over a randomly selected area and record plants found at each one-foot or one-meter "point" or interval, for example (Weller and Fredrickson 1974). Others have used the spokes of a wheel or a frame set up at intervals with spear-points that touch the vegetation. The major concept underlying these methods of collecting data is the randomness of the chance to encounter various plants, so that the abundant ones should be recorded most often and in proportion to less abundant species. This system also can be used to show the distribution of open water, water depth, and types and location of submergent plants.

The line-intercept method is similar, but all plants touching one side of a tape are tallied. Numbers, therefore, are larger and distribution is better shown. Often this method is combined with point-counts at set intervals, to quantify records of water depth, submergents, or open water.

Small quadrats are especially favored where detailed plant observations are to be made. One can set these out randomly by gridding an area, numbering the intersections or plots, and using a statistical table or a calculator that generates a series of random numbers for quadrat selection. Numerous less-structured systems can be found. For statistical measurement of variation, three samples are minimal, but this small sampling is unlikely to achieve a significant level of population assessment. The number of plots needed usually may be assessed grossly by finding the number at which measured changes are no longer significant (termed a species-number plot). If, for example, you do twelve quadrats, and the number of species found and their distribution do not change from numbers eight through twelve, chances are good that a sample size of eight or ten will adequately document what you want to show. Obviously, this sample size varies with the number of species present, their spatial distribution, and their numerical abundance—especially for the rarer species. The same technique can be applied to quadrat size by experimenting with different sizes of plots to assess their advantages for measuring number of species or other information.

Quadrats of rectangular shape seem to be preferred by many plant ecologists. Data recorded in the quadrat can vary from clipping and counting to detailed cover mapping. In some cases, *exclosures* are used to keep herbivores out of the plots, allowing the assessment of the amount of cover consumed by muskrats or birds. Relative abundance data may be gained on cover (a real distribution of plants), by ranking in percent or numbers; detailed information on procedures should be obtained from a current ecology text.

A final note about plants: Plant life form, rather than details of species, may be all that is necessary for some studies. In such cases, the mapping of tall versus short emergents, or distribution of shrubs versus emergents, or submergents versus emergents, may be less demanding than determining all species, and yet give a measure of available habitat. However, the procedures are the same.

Water and Water Characteristics

The simplest and probably the most important single measure of water at a study site can be done with a staff gauge. By using a simple vertical meter stick, one can make a series of obervations on presence or absence of water at a particular site over time (hydroperiod), its depth, and its seasonality. This dictates not only the plant community, but also what animals one finds when and where.

The study of chemical or physical characteristics of water is much more complex. There are available now some simple test kits to measure certain parameters of pollution, but most amateurs should leave that effort to professionals. There are some excellent limnological texts, however, that provide a good background for comprehension of the patterns and processes that occur in water (Wetzel 1975). Chemical analysis of dissolved solids in the water is of particular interest because there tends to be a correlation between high concentrations and high biological productivity. These concentrations usually are measured electrically by specific conductance and require a fairly expensive instrument. However, some features of the water can be measured simply, such as water clarity and pH. Water turbidity is of special interest in marshes because they tend to be clear, and intrusions of silt or of nutrients that cause algal blooms are conspicuous. Such turbidity stresses and may kill submergent plants. A weighted, patterned disk tied to a measuring rope can be lowered into the water until the pattern is obscured. The measured depth is an index to clarity, and microscopic examination may reveal the cause of turbidity. Although the hydrogen ion concentration (pH) may not be useful information except where it is extreme, it is easily measured with pH paper available from biological supply houses. Most inland, freshwater marshes range from near normal (6.5–7.0) to alkaline (8.0–8.5). Highly acid bogs may be 4.0–5.5.

Invertebrate Richness and Abundance

The diversity and abundance of the larger forms visible to the naked eye (macroscopic) such as insects and crustaceans, and of the tiny (microscopic) protozoans and other less well known groups of invertebrates, **tend** to discourage amateur study. But the larger and more common forms can be sorted and easily placed in large taxonomic categories (Pennak 1978), providing the opportunity to see the diversity and abundance of life often found in seemingly sterile waters, and perhaps explain why certain ducks or shorebirds feed where they do.

Swimming invertebrates usually are sampled by nets or traps. For very small planktonic forms, fine-meshed sweep nets or dragnets are preferred. They can be simulated by cheesecloth, plastic mesh, or fine metal screen, but still finer nets of durable quality must be purchased at biological supply houses. Dip nets with modest-sized openings (twenty meshes per inch or eight per centimeter) let water and silt pass through, while trapping the larger, more visible invertebrates of greatest interest to the amateur (and especially to fish, amphibians, and birds). Minnow traps, available in sporting-goods stores, also work for dragonflies and other aquatic nymphs, and for swimming adults. Dragonflies, damselflies, midges, caddis flies, and mosquitoes, which leave the water as flying adults, can be quantified in small cages or emergence traps built over the water, which hold those insects that have emerged in that surface area.

Many of the larger invertebrates of marshes are bottom or benthic forms that can be sampled by various dredges and corers (Paterson and Fernando 1971). Some precise and expensive samplers designed specifically for bottom sampling work well in mud, sand, or even gravel bottom (Kendeigh 1961) but often are useless in marshes because of the plant debris and root stalks. Corers can be made of plastic drain pipe or tubing such as those used in plumbing. When the edge is sharpened, they can be forced into the substrate by rotating as they are pushed. If plugged with a cork or the standard screw top, the suction holds the debris in the tube as it is quickly transferred to a bucket. Such dredge materials then must be washed through sampling screens and the remains sorted. This is a slow but exciting process, revealing the variety and abundance of invertebrate life.

Flying insects can be taken by "butterfly" net, or by "sticky-trap"—an adhesive sprayed on boards of specific sizes where landing insects are caught.

Remember that some invertebrates are endangered, so some states do not allow trapping and sampling of invertebrates without permit. Check the local laws!

Birds

Birds have been among the most interesting animals of marshes because of their variety, population density, colors, and conspicuous behavior. Armed with binoculars and a bird guide, one can assess the number of species present (species richness) or distribution by vegetative type fairly easily, but determining population size and density per habitat type is much more difficult. Because of legal protection, difficulty of finding nests, and concern for disturbance of nesting birds, the census of birds often is best directed to counts of singing males on perches (blackbirds, sparrows, wrens), loafing sites (ducks, geese), or feeding pools (grebes, coots). However, professionals find that some species are best counted by their nests because the adults are so inconspicuous (bitterns and grebes), or occur in such large colonies of densely packed nests (gulls, terns, egrets, herons, ibises) that they cannot be estimated properly from a distance. This is a delicate job and timing is important, so only experienced and official biologists should attempt such censuses; the damage done in one survey could be greater than the benefits to the conservation of the species contributed by the whole project. Where large marshes or cover types cannot be censused, blocks or quadrats (Figure 16), strip transects, or even line transects can be used as sampling units to compare two areas or habitats.

Mammals

Marsh–dwelling mammals are extremely difficult to census, and trapping is complicated by legal problems and permits, by techniques, and by difficulties of data interpretation that prevent the beginner from doing meaningful work. Local data on trappers' catches of mink, weasels, and muskrats are extremely useful population measures, but are of unknown quality. Indexes to mammal use, such as numbers of muskrat lodges or the incidence of mouse or muskrat clippings, provide some crude measure when variation in populations is extreme. In large areas these are best sampled by using intercepting transects or quadrats.

Amphibians and Reptiles

Although qualitative sampling for species identification, age, or stage of development can be done by hand grab or by dip net, quantitative data are difficult to obtain. The most successful efforts have been with snakes and walking amphibians trapped by drift fence and can traps as they move from area to area. The numbers of tiger salamanders taken during spring breeding movements have been unbelievable at certain sites.

Fish

Even though game species are rarely involved, state conservation agencies restrict fish sampling by special license for scientific work. Laws differ from state to state and must be checked before any capture occurs. Netting is the usual means of capturing fish in ponds or lakes, but nets used in marshes are fouled by algae or vegetation. Dip nets occasionally catch fish, but provide no systematic approach or quantitative data. Minnow traps work for some species but not for others. Selective poisons and electric shocking have been used by professionals, but they may be dangerous, and all require special permits. Moreover, the data are difficult even for professionals to evaluate. Depending on the purposes of study, harvest data on game fish or commercial harvest may provide good data on local areas.

APPENDIX B
Marsh Management Techniques

Waterfowl management areas include marshes and several other kinds of palustrine wetlands such as meadows, deep marsh, shrub, and forested swamps. Many of these wetland types contain some marsh vegetation important to waterbirds, mammals, and other vertebrates for food, and for protection from weather and predators. The most common categories of freshwater wetland management areas are (1) moist-soil impoundments, emphasizing the development of shallow-water meadow plants productive of seeds used by waterfowl in the nonbreeding period (Fredrickson and Taylor 1982); (2) deep emergent marshes, favoring more permanent perennial plants used as nesting areas for birds, food by muskrats and nutria, and cover by hunters; and (3) bottomland floodplain areas called "greentree reservoirs" where food resources such as acorns and other foods from shrubs and tree (collectively termed "mast") are utilized. Because of the focus of this book, I will emphasize the first two categories and encourage those interested in greentree impoundments to read a newly available sourcebook (Fredrickson and Batema 1992).

Wildlife managers and fisheries managers have been involved in wetland management and enhancement efforts for many years, and often are the best source of wetland restoration techniques that have been tried and work in a certain region. Most of the techniques for habitat management are based on natural processes in wetland systems, and thus also influence other wetland values and functions. However, few of these practices have been subjected to long-term experimental testing and evaluation. Much of this material has been published, and there now are several summaries available (Kusler and Kentula 1990, Payne 1992, U.S. Soil Conservation Service 1992) that cover commonly managed wetland types. This appendix brings together some generalizations and a few examples of management procedures that have been standardized and that have proved generally reliable (see Weller 1989b). From a wildlife standpoint, such efforts have emphasized waterfowl and furbearers; very little work has dealt with other groups of birds, although some papers reflect that successful efforts on game species also may favor nongame species or watchable wildlife. These standardized management procedures can be used equally well to enhance other wetland functions, such as reducing turbidity, protecting shorelines from erosion, and maintaining esthetic values, and are applicable to the special situations of restoring, enhancing, or creating wetlands to meet mitigation requirements.

Restoration of drained wetlands has been common where there was an opportunity to reclaim a major wetland of high wildlife value that has been degraded. This commonly involves a failed drainage project where land values have declined, so that federal or state agencies can purchase the land and attempt restoration. Numerous National Wildlife Refuges such as Tule Lake in California and Agassiz Lake in Minnesota fall into this category (Laycock 1965). Still older examples are available in the European literature (Fog 1980). Enhancement of wetlands involves an attempt to improve the wildlife values of a wetland that has not been drastically perturbed, but that managers believe could be producing wildlife at a higher level more of the time. Wetland creation has been a less common practice among wildlifers, but the conversion of terrestrial communities to wetlands at the margins or upper reaches of reservoirs may result from creating impoundments for other purposes such as water supply or flood control. Additionally, mining operations such as gravel removal may result in suitable wildlife habitats, and considerable information is available (Svedarsky and Crawford 1982).

Goals and Objectives

Philosophically, many of the goals of wildlife managers overlap with those of other wetland managers and spring from the desire to preserve natural landscape units and functional values. Managers differ, however, in whether they prefer to use artificial methods that may give more immediate results, versus more natural processes that tend to take longer but may be less expensive and longer lasting (Weller 1978a). Benefit–cost ratio strongly influences the choice of strategy as some functional but very expensive techniques such as hand transplants can be more easily justified in mitigation procedures than in conservation efforts for wildlife. Short-term goals often have been the driving force in habitat management of wetlands for wildlife. Such goals have been directed toward 1) increased production of harvestable wildlife for hunting through the creation of more and better nesting sites, increased food supplies, more resting and roosting places, and reduced predation; and 2) improved conditions for hunters in pursuit of wildlife by providing cover patches or blind sites and by facilitating access to wetland areas for hunting by means of roads and boat channels.

Many wetland wildlife managers cherish the rich natural values and esthetic aspects of the natural system. They foster a natural management philosophy as a code of ethics (Errington 1957), and are dedicated to preserving a naturally functioning ecosystem in perpetuity (Weller 1978a, 1989b) and to maintaining or adding biological diversity (Sanderson 1974, Mathisen 1985, Rundel and Fredrickson 1981). Other managers place first priority on maximal production of the targeted game species, but without intent of opposing other natural values and functions. With the decline of wetlands in the United States, these approaches should be incorporated into a unified plan favoring all wetland functions, regardless of funding source or project goal. The consideration of multiple interests and approaches has been enhanced by the use of public hearings of local residents, laypersons, and experts.

Preconstruction Considerations

Geomorphology, Landscape, Patch Size, and Patterns. It has become increasingly apparent that waterfowl and other migratory birds, as well as amphibians, reptiles, mammals, and sometimes even wetland fish, do not satisfy all their needs in one wetland. This is particularly true of the breeding period when very specific requirements for foods or nest sites may exist in comparison to postbreeding and its activities such as migration or wintering. Thus, wetland diversity and density may figure prominently in satisfying these needs, a consideration that may not be met with the restoration or creation of only one wetland, when a complex may be necessary. Wetland complexes have been recognized as important by many workers, and data presented by Brown and Dinsmore (1986, 1991) suggest that complexes increase species richness over solitary wetlands of similar size, but that size alone is an important feature. General studies of wetland density in prairie pothole habitats, where water cycles influence wetland numbers from year to year, show general correlations between wetlands and waterfowl abundance (Leitch and Kaminski 1985). The diversity of wetland types also is deemed vital as losses of small wetlands in drought years have a particularly great impact on certain species (Evans and Black 1956). These smaller areas are especially vulnerable to some delineation systems favored by land-development interests.

Larger wetlands are known to provide greater numbers of habitats, and therefore are more likely to attract a greater number of species (i.e., species richness) (Brown and Dinsmore 1986). For these reasons, there is a tendency to acquire and restore large units, but it is also recognized that small areas may provide specialized habitats for certain species, and management for those may require size considerations (Evans and Black 1956).

Configuration (e.g., length of shoreline in relation to area of the wetland) seems to be an important influence on numbers of territorial species that an area can support, but there are little substantiating data (Mack and Flake 1980, Kaminski and Prince 1984). Some workers have suggested that other vegetation patterns are more influential than shoreline length (Patterson 1976). Managers often successfully create artificial wetlands with complex configurations to provide isolation for breeding pairs.

Designing for Water Depth. Preconstruction planning involving detailed contour mapping of prospective sites is essential (Verry 1985). Site observations during natural flooding periods also are useful because contour maps may not provide the precision essential in water-level regulation. For example, large-scale contour intervals are usual on construction maps, when the ultimate precision required for water-level control may be a matter of inches or centimeters.

Contouring with earth-moving equipment is commonplace, and should be used to create water depths associated with the desired plant community. Islands, bays, and other structural features can be created during construction if soil character and shoreline protection are considered. Where such work is done on areas with a rich seed bank, soil should be moved off site and returned as topsoil for the merits of both its organic content and seed bank. Dams and levees often are built of borrowed soil taken from the water side to create some deep-water sites or to facilitate installation of water control structures. When basin shape is modified by scraping, a barren substrate may be created (Kelting and Penfound 1950) that must be reseeded or await the natural processes of seed transport, germination, and local seed production. Returning topsoil also may reduce invasions by exotics where they are an issue.

Regulating Runoff, Erosion, and Turbidity. Rainfall-evaporation patterns and watershed-wetland size ratios are of special importance in selection of impoundment sites (Verry 1985). Special considerations are necessary where uplands have been modified by land-use practices. Farming may increase the mean annual silt load and thereby eutrophication. Intensive grazing on the waterway slopes can increase peak surges of water entering the wetland during storm events, washing out vegetation and water-control structures. Urban development may increase runoff resulting from parking lots, roadways, and roofs. In wetland restoration or enhancement projects, water-control structures must be carefully engineered to handle these added burdens. Additionally, upslope protection can be achieved through water diversions to streams, grass plantings to absorb more rainfall, or smaller impoundments that serve as catch basins for both water and silt.

These solutions also are relevant where wetlands are created from terrestrial sites, in which case site selection is extremely crucial and requires knowledge of the slope, area, and rainfall data of the watershed (Verry 1985). Storm events always seem to exceed runoff projections and are particularly damaging in the early stages of wetland development.

In any wetland management program, modified water levels, exposed banks, and unvegetated bottom are vulnerable to wave and current action. Decreased wetland productivity may result from erosion of shorelines, elimination of vegetation, and increased water turbidity. Importing firmer soil, gravel, or rock may be necessary, and prepared riprap can be used in extreme cases of erosion. Two steps that can be taken to prevent erosion: are exposing the shoreline by dewatering (drawdown) until vegetation has become established, and delaying flooding until the bottom has been stabilized with emergent and preferably submergent vegetation. These steps also will reduce turbidity of the water that may prevent vegetation establishment when reflooding occurs.

Management Procedures by Process and Proximate Objective

Managing the Seed Bank. Although the longevity and abundance of seeds of wetland plants have been known at least since the 1930s (Billington 1938), and although early marsh managers advised location and utilization of sites with a natural seed bank (Addy and MacNamara 1948, Crail 1955, Singleton 1951), managers have varied by time and place as to how well they have utilized this general principle. Seeding of Japanese millet or use of native millets and smartweeds for seed sources was widespread among National Wildlife Refuges in the 1940s and 1950s (Linduska 1964). More recently, moist-soil management and other types of strategies that use natural seed banks have become the accepted standard of a generation of managers who favor natural processes and minimal expenditures on machinery and workforce. Management of both water depth and hydroperiod is the major strategy employed (Fredrickson and Taylor 1982, Rundel and Fredrickson 1981), and the techniques are now more widely recognized and appreciated.

Despite their common longevity, seed banks may be limited in diversity or even nonexistent in situations of long inundation where wetland emergents have not existed for many years; aquatic plants or long-lived terrestrial plant seeds may survive, but species of value to the wetland restoration may not occur (Pederson and van der Valk 1984). Where available, use of soil from local wetlands may be useful in resolving this problem.

Managing Plant Succession through Water-Level Manipulation. Because it has long been recognized that hydroperiod and water depth dictate composition, density, and growth of plant species, waterfowl managers use water not only to satisfy physiological requirements of the seed for germination, but also to influence rates of growth, tuber production, and general plant productivity (Beule 1979). To provide the water conditions that induce seed germination and plant growth, most wetlands created or modified for wildlife make provision for complete dewatering by use of control structures (Atlantic Flyway Council 1972, Payne 1992). The latter is a vital consideration if any influence over plant community is desired, and the engineer should be alerted to the need of total dewatering. Subsequent modifications of dike height must consider water-control structures and the potential for dam failure or erosion. Strategies for achieving the desired goals will be outlined under examples of several wetland types that have been commonly and successfully managed.

The most common method of deepening a natural or restored wetland for waterfowl has been to install a dam or dike to impound more water, and to incorporate a water-control structure that allows dewatering. Increasing water depths in basins that do not have an adequate supply can best be accomplished most economically by gravity-flow systems such as stream diversions and upslope reservoirs. A more reliable but expensive alternative is a well and pump, but this is a perpetual expense, and may be a source of disagreement among different user groups. Modifying shallow wetlands to create open pools and deeper water can be done by dewatering and a bulldozer, by drag-line movement of basin substrate, or by blasting with explosives (Strohmeyer and Fredrickson 1967, Mathiak 1965).

Regulating Vegetation by Herbivore and Fire. Wetlands, especially those dominated by persistent perennial emergent plants, are renowned for their attractiveness to the native muskrat throughout much of North American and the introduced nutria in the South. Overpopulations resulting in eat-outs are well documented in northern and midwestern marshes (Errington et al. 1963), and dramatic population changes also exist in eastern brackish marshes (Dozier 1947) and coastal marshes (Lynch et al. 1947, Palmisano 1973). Ultimately, the area is denuded and populations of other wildlife are drastically impacted (Weller and Spatcher 1965). Managers often are unprepared for this event and may lack methods for control because of harvest regulations. In large areas, control may be impossible due to size and logistics of trapping alone. The resultant open water may persist for many years unless dewatering is used to induce revegetation.

Beavers likewise can impact willows, cottonwoods, and other highly palatable plants at marsh edges, whether the plant is used only for food or for lodges and dams as well (Beard 1953). Flooding by beavers of other terrestrial or wetland vegetation often is regarded as serious by managers not only because of plant mortality but because water-level stability within a wetland may not be conducive to maximal community diversity and productivity of many wetland species.

Livestock such as cattle, sheep, and goats can be useful management tools, but such grazing may be difficult to regulate because of public pressure for grazing rights or because of our lack of understanding of the carrying capacity of wetlands under variable and often uncontrollable conditions. Grazing exposes tubers then utilized by grubbing geese such as snow geese in both southern and eastern coastal marshes (Glazener 1946), but also can eliminate favored duck food plants (Whyte and Silvy 1981). Fencing is the simple tool used by managers all over the world to change the character of overgrazed wetlands (Fog 1980), but better management may involve regulation of grazing level and not merely total exclusion.

Fire has been used by farmers and ranchers for years to increase forage and hay crops, but it can be devastating to nesting ducks and other birds (Cartwright 1942). The response of wintering geese to fire has resulted in a policy of periodic burning on refuges to reduce vegetation and to expose tubers for grubbing geese (Lynch 1941). Fire also has been used in northern marshes with peat bottoms to create deep-water openings, but control of the fire often is difficult (Linde 1985). Considerable work has been done recently to explore the role of fire in marsh succession (Smith 1985a, 1985b), and clearly much more of this type of work is needed in all vegetation types (Kantrud 1986).

Controlling Weeds. Part of maintaining a wetland that is attractive to wildlife requires a balance of open water and vegetation. Because nesting birds and migratory waterfowl especially are influenced by these conditions, considerable effort goes into creating suitable habitat.

Some of the techniques such as drawdown create suitable seeding situations for weed species. Undesirable plants, particularly willow and cattail, can be extremely troublesome as they outcompete annuals and eliminate openings favored by ducks. The capability of drying out the area, or of flooding it to excessive depths, may allow control of some nuisance species (Fredrickson and Taylor 1982).

Management of Palustrine Persistent Emergent Wetlands: The Deep Marsh

As discussed in chapter 8, deep marshes are used especially by nesting waterbirds and other marsh wildlife. Most managers prefer to have not only well-designed water-control structures but a reliable source of water to use in modifying water depths to manage plants. A diversity of plants of various life forms is preferred to serve various animals: Marginal nonpersistent emergent plants produce large seed crops for songbirds, rails, and ducks; deeper water persistent emergents are excellent for nest sites and provide tubers useful to herbivores; and submergent plants often provide food directly or serve as substrates for invertebrates.

Water depths dictate dynamic vegetation patterns in wetlands subject to seasonal and annual variation in water supply. They may vary from lakelike aquatic conditions to near-terrestrial vegetation due to hydrologic perturbations. Wildlife respond directly and vary greatly in species richness and population abundance (Weller and Spatcher 1965).

A well-established technique to reestablish vegetation after it has been eliminated by high water or muskrat eat-out is to dewater the area by use of the water-control structure or pumping. Germination from the natural seed bank (van der

Valk and Davis 1978) provides most of the source of plants, but enhanced production of persistent plants as a result of tubers and rhizomes also results (Weller 1975a). The drawdown–revegetation cycle is commonly practiced by conservation agencies in many midwestern states (Linde 1969) and in situations where time is less important relative to costs. This method seems to simulate natural processes and events without lasting damage. Even fish populations of those areas seem responsive and pioneering.

Reestablishment of vegetation may result in excessively dense vegetation or plant species not suitable for the intended wildlife, whereupon the natural herbivores may move in and create suitable openings and modify species composition. Where such herbivores are absent, cover-to-water ratios can be modified with fire (Kantrud 1986), grazing (Kirsch 1969), or more direct (artificial and costly) methods such as chemical sprays (Martin et al. 1957, Beule 1979) or mechanical destruction by roller or cutting (in or out of water) (Nelson and Dietz 1966, Linde 1969).

Moist-Soil Impoundments for Nonpersistent Emergents: The Shallow and Temporary Marshes

Migratory waterfowl feed less on high-protein animal foods in fall and winter and instead use seeds or foliage. Seed production is especially enhanced by creating or maintaining shallow-water areas where nonpersistent emergents such as millets, smartweeds, spike rushes, and other marsh-edge plants germinate and produce seed (Fredrickson and Taylor 1982). The usual procedure is to construct low dams equipped with a simple structure for water-level control, and sometimes levees to direct or enclose water areas. Such areas produce mud flats typical of any marsh drawdown, and are designed for feeding areas for waterfowl rather than for breeding (although they may be suitable for some rails and songbirds). Drawdowns are performed in summer growing periods, but the timing depends on the latitude. One must allow sufficient drying so that annual seeds will be produced, but growth of weedy species will be avoided. Because of the climatic regime in southerly areas, both spring and late summer crops may be produced where water is available and levels can be controlled. In rice areas, low-level terraces that are opened and closed by machinery work quite well, but the intended crop of annuals must determine the structure design and water depth.

Areas also are flooded in spring or fall and dewatered gradually to create mudflat conditions attractive to migrant shorebirds and certain ducks (e.g., green-winged teal) that feed on invertebrates (Fredrickson and Taylor 1982, Rundel and Fredrickson 1981). Too-rapid drying produces more terrestrial species (Harris and Marshall 1963), so soil conditions and water-level regulation are extremely important. The retention of some moisture on the flat is essential to ensure full maturity of seeds and germination of late-maturing plants such as smartweeds. Here, as in any drawdown, a thunderstorm can produce flooding and the loss of a year's crop. Typically, such areas are reflooded in the fall prior to the arrival of waterfowl and other migrants. Flooding to make food available to migrant waterfowl involves regulation of the water-control structure (except in cases of high rainfall and flow-through rates) to maintain depths of six to eighteen inches (fifteen to forty-six centimeters) so that birds can swim but still dabble and tip up for food. Dabbling ducks such as green-winged and blue-winged teal, mallards, and pintails find ideal food and water conditions in such situations. Deeper areas may be utilized by shallow divers such as the ring-necked ducks. Ideally, several different wetland management units should be available to provide a wetland complex with different water depths and foods (Fredrickson and Taylor 1982).

APPENDIX C
Scientific Names of Plants and Animals Mentioned in the Text

Alligator *Alligator mississippiensis*
Arrowhead *Sagittaria* spp.
Backswimmers (Notonectidae)
Bass, Largemouth *Micropterus salmoides*
Beaver *Castor canadensis*
Beetle, Predaceous Diving (Dytiscidae)
Beetle, Whirligig (Gyrinidae)
Beggar'stick *Bidens* spp.
Bison *Bison bison*
Bittern, American *Botaurus lentiginosus*
Bittern, Least *Ixobrychus exilis*
Blackbird, Red-winged *Agelaius phoeniceus*
Blackbird, Yellow-headed *X. xanthocephalus*
Bladderwort *Utricularia* spp.
Bobolink *Dolichonyx oryzivorus*
Bug, Giant Water (Belostomidae)
Bullfrog *Rana catesbeiana*
Bullhead, Black *Ictalurus melas*
Bulrush, Alkali *Scirpus paludosus*
Bulrush, Hardstem *S. acutus*
Bulrush, River *S. fluviatilis*
Bulrush, Softstem *S. validus*

Burreed *Sparganium* spp.
Cactus, Prickly Pear *Opuntia* spp.
Caddis flies (Tricoptera)
Canvasback *Aythya valisineria*
Carp *Cyprinus carpio*
Carp, Grass *Ctenopharyngodon idella*
Cattail *Typha* spp.
Cattail, Broadleaf or Common *T. latifolia*
Cattail, Hybrid *T. glauca*
Cattail, Narrowleaf *T. angustifolia*
Cattail, Southern *T. dominguensis*
Celery, Wild *Vallisneria americana*
Chufa or Flatsedge *Cyperus esculentus*
Clam, Fingernail *Sphaerium* spp.
Coot, American *Fulica americana*
Copepod, e.g., *Cyclops* spp. (Copepoda)
Cottonwood *Populus deltoides*
Crane, Sandhill *Grus canadensis*
Crane, Whooping *G. americana*
Crane flies (Tipulidae)
Crayfish (Decapoda)
Curlew, Long-billed *Numenius americana*

Cutgrass, Rice *Leersia oryzoides*
Cyclops (Copepoda)
Cypress, Bald *Taxodium distichum*
Damselflies (Odonata)
Daphnia or Water Fleas (Cladocera)
Dove, Mourning *Zenaidura macroura*
Dragonflies (Odonata)
Duck, Dabbling *Anas* spp.
Duck, Fulvous Whistling *Dendrocygna bicolor*
Duck, Inland Diving *Aythya* spp.
Duck, Ring-necked *Aythya collaris*
Duck, Ruddy *Oxyura jamaicensis*
Duck, Sea (Tribe Merginini)
Duck, Wood *Aix sponsa*
Duckweed, Lesser *Lemna minor*
Duckweed, Star *L. trisulca*
Egret, Cattle *Bubulcus ibis*
Eider *Somateria* spp.
Flatsedge or Chufa *Cyperus esculentus*
Fox, Red *Vulpes fulva*
Frog, Leopard *Rana pipiens*
Gadwall *Anas strepera*
Goose, Canada *Branta canadensis*
Goose, Snow *Chen caerulescens*
Grass, Whitetop *Scholochloa festucacea*
Grebe, Eared *Podiceps nigricollis*
Grebe, Horned *P. auritus*
Grebe, Pied-billed *Podilymbus podiceps*
Grebe, Red-necked *Podiceps grisegena*
Grebe, Western *Aechmophorus occidentalis*
Gull, Franklin's *Larus pipixcan*
Harrier, Northern *Circus cyaneus*
Heron, Black-crowned Night *N. nycticorax*
Heron, Great Blue *Ardea herodias*
Heron, Little Blue *Egretta caerulea*
Heron, Tricolored *Egretta tricolor*
Hyacinth, Water- *Eichhornia crassipes*
Hydrilla *Hydrilla verticillata*
Ibis, White-faced *Plegadis chihi*

Jellyfish, Freshwater *Craspedacusta sowerbyi*
Jewelweed *Impatiens* ssp.
Juncus or Rush *Juncus* spp.
Killdeer *Charadrius vociferus*
Kite, Snail or Everglade *Rostramus sociablis*
Lark, Horned *Eremophila alpestris*
Leech (Hirudinea)
Loon, Common *Gavia immer*
Mallard *Anas platyrhynchos*
Mayflies (Ephemeroptera)
Meadowlark *Sturnella* spp.
Merganser *Mergus* spp.
Midge or Marshflies (Chironomidae)
Millet or Barnyardgrass *Echinochloa* spp.
Mink *Mustela vision*
Minnow, Fathead *Pimephales promelas*
Moose *Alces americana*
Moorhen, Common *Gallinula chloropus*
Mosquito (Culicidae)
Mouse, Meadow *Microtus pennsylvanicus*
Muskrat *Ondatra zibethicus*
Muskrat, Round-tailed *Neofiber alleni*
Nutria *Myocastor coypus*
Oriole, Northern *Icterus galbula*
Ostracods or Seed Shrimp, e.g., *Candona* spp. (Ostracoda)
Otter *Lutra canadensis*
Owl, Snowy *Nyctea scandiaca*
Pheasant, Ring-necked *Phasianus colchicus*
Pike, Northern *Esox niger*
Pintail, Northern *Anas acuta*
Pondweed *Potamogeton* spp.
Pondweed, Sago *P. pectinatus*
Potato, Duck *Sagittaria* spp.
Ptarmigan, Willow *Lagopus lagopus*
Raccoon *Procyon lotor*
Rails (Rallidae)

Rail, King *Rallus elegans*
Rail, Sora *Porzana carolina*
Rail, Virgina *Rallus limicola*
Rail, Yuma Clapper *Rallus longirostris yumanensis*
Rat, Rice *Oryzomys palustris*
Redhead *Aythya americana*
Reed, Common *Phragmites communis*
Rush *Juncus* spp.
Salamander, Tiger *Ambystoma tigrinum*
Scuds or Sideswimmers (Amphipoda)
Sedge *Carex* spp.
Shoveler, Northern *Anas clypeata*
Shrew, Short-tailed *Blarina brevicauda*
Shrimp, Fairy *Branchinecta* spp.
Shrimp, Seed (Ostracoda)
Sideswimmers or Scuds (Amphipoda)
Skunk, Striped *Mephitis mephitis*
Smartweeds *Polygonum* spp.
Snake, Garter *Thamnophis* spp.
Sora *Porzana carolina*
Sowbugs (Isopoda)
Sparrow, Song *Melospiza melodia*
Sparrow, Swamp *M. georgiana*
Spikerush *Eleocharis* spp.
Sponge, Freshwater (Spongillidae)
Stickleback, Brook *Culaea inconstans*
Stork, Wood *Mycteria americana*
Strider, Water (Gerridae)
Swallow (Hirundinidae)
Swan, Trumpeter *Cygnus buccinator*

Teal, Blue-winged *Anas discors*
Teal, Green-winged *A. crecca*
Tern, Black *Chlidonias niger*
Tern, Common *Sterna hirundo*
Tern, Forster's *S. forsteri*
Toad, American *Bufo americanus*
Toad, Houston *B. houstonensis*
Turtle, Blanding's *Emydoidea blandingii*
Turtle, Mud *Kinosternon* spp.
Turtle, Painted *Chrysemys picta*
Turtle, Snapping *Chelydra serpentina*
Water-crowfoot *Ranunculus* spp.
Water fleas, e.g., *Daphnia* spp. (Cladocera)
Water-hyacinth *Eichhornia crassipes*
Water Lily, Yellow *Nuphar advena*
Watermeal *Wolfia* spp.
Water Milfoil *Myriophyllum* spp.
Water Rat or Round-tailed Muskrat *Neofiber alleni*
Weasel, Least *Mustela rixosa*
Weasel, Short-tailed *Mustela erminea*
Whitetop *Scolochloa festucacea*
Wigeongrass *Ruppia maritima*
Willow *Salix* spp.
Wren, Marsh or Long-billed *Cistothorus palustris*
Wren, Sedge or Short-billed *C. platensis*
Yellowthroat *Geothlypis trichas*

APPENDIX D
Glossary of Terms Used in the Text

Above-ground biomass. Plant materials such as leaves, stems, and flowers found above the soil or basin level as opposed to that below.

Adaptation. Genetically based modifications in anatomy, physiology, and behavior that better equip an organism to deal with its environment.

Alkali lakes. Bodies of water located in areas where a high rate of evaporation concentrates salts such as sulfates and chlorides.

Basin. Topography of the land that creates a depression that holds water.

Below-ground biomass. Plant materials such as roots, rhizomes, and tubers found beneath the soil or basin surface.

Benthos. Organisms such as clams that use the bottom of the marsh or lake, and that sometimes are imbedded in the substrate soil and organic matter.

Biomass. The total weight of living material (plant and animal) found on a site or sample area.

Capillary action. A physical action produced by surface tension that draws water upward through fine pores in the soil in opposition to gravitational forces.

Competition. The negative interaction of two organisms seeking to use the same resource, such as food or habitat.

Cover-to-water ratio. The percentage of cover in relation to the total of open water: e.g., fifty-fifty equals half of each.

Decomposition. The combined processes by which organic matter is broken down from its original form to its organic and inorganic constituents by leeching, detritivores, bacteria, and fungal decomposers. Some ecologists restrict the use of this term to action by decomposer organisms.

Detritivore. Organisms that utilize dead organic tissue, such as plant material, as food, and break it into smaller and smaller pieces.

Discharge zone. Underground water that comes to the surface on hillsides or wetlands due to intersection of the water table.

Diversity. Implies the variety of life, usually viewed positively by wildlife managers.

Drawdown. The (usually) intentional lowering of water levels for the purpose of allowing germination of seeds in the substrate. Sometimes used to describe conditions resulting from a drought.

Ecology. The science that studies organisms in relation to their environment and attempts to explain patterns and processes that make biological systems work.

Ecosystem. A community of living organisms and their physical environment that functions as an entity in energy flow and nutrient cycling.

Ecotone. A term once commonly used to describe the edge between major plant communities, e.g., grassland and forest. Some employ the term to describe any vegetational edge.

Emergent. A plant that grows rooted in shallow water but the bulk of which emerges from the water and stands vertically. Usually applied to herbaceous rather than woody vegetation.

Endangered species. A species of animal or plant so rare that it is given this classification on a special list by the U.S. Fish and Wildlife Service, and it and its habitat are entitled to federal legal protection.

Environmental impact statement. A document prepared to outline and justify an action such as construction that will modify the environment substantially. Often includes preventive or compensatory action.

Eutrophication. The process of gradual enrichment that occurs in a body of water through natural inflow and accumulation of nutrients, or indirectly through human action such as fertilization in upland grass or crops.

Evaluate. The qualitative or quantitative process of assessing the value of a habitat for wildlife. Currently done by most federal agencies using the U.S. Fish and Wildlife Service's Habitat Evaluation Procedure (HEP) or the U.S. Army Corps of Engineers' Wetland Evaluation Technique (WET).

Everglades. A unique wetland type in southern Florida made up of communities of aquatic grasses and sedges interspersed with hammocks of tropical trees.

Evolution. The genetic process of change that occurs in species with time, resulting from natural selection and survival of those best adapted to the situation.

Exotic. An alien organism from a foreign country that is released intentionally or accidentally in another country and that survives. Often very successful as a species but may compete with native forms.

Fen. A unique and localized sedge-moss wetland produced where slightly alkaline water emerges at the surface. Bogs have similar types of vegetation but tend to be acid.

Floating-leaf plant. An aquatic plant with large leaves that float on the surface of the water and that are attached by flexible stems to roots and tubers in the substrate.

Floating plant. An aquatic plant that floats on the surface or occasionally in the water column and that typically has roots that extract nutrients from the water.

Function. Processes that go on within a biological system and have values to that system or to humans. Production of organic matter, for example, not only enriches the marsh but provides potential food for humans or for organisms that humans utilize.

Furbearers. A term used to describe diverse mammals that are trapped for their pelts and have commercial value, such as muskrats and mink.

Germination. The process of embryo development and growth of a seedling plant from a seed, usually induced by a specific set of moisture, temperature, and chemical conditions.

Glacial. Reference to any product or action of glaciers such as the sheet glaciers that once covered northern North America and created myriads of marshes in the prairie region of the United States and Canada.

Great Basin. An intermountain region of the western United States characterized by low rainfall, high evaporation, and highly saline lakes. Water supplies to such marshes often originate from snowmelt in the mountains.

Habitat. The place where an animal (or plant) resides and finds food, water, cover, and space.

Habitat diversity. A qualitative statement of the structure of the vegetation that forms potential habitats for wildlife. Generally speaking, the more diverse the structure, the more potential habitats exist for diverse wildlife. Usually such diversity is one of the objectives of wildlife management.

HEP. An acronym for the U.S. Fish and Wildlife Service's Habitat Evaluation Procedure.

Hydric soil. Soils with deposits of organic matter and minerals characteristic of anaerobic conditions suggesting development in wetlands.

Hydrology. The study of water movement on the earth's surface and in the underlying soil and rocks.

Hydroperiod. The duration of flooding or water presence in a wetland. This usually dictates what plants germinate and grow.

Hydrophytes. Water-adapted plants found where water is at or near the surface, forming good indicators of wetlands.

Insolation. Absorbtion of the sunlight on the earth or water's surface.

Instability. Variation in physical factors such as water depth that influence and even regulate biological processes such as plant growth.

Irrigation. A system devised to capture and distribute water in suitable amounts and timing for agricultural crops.

Limnology. The science that deals with the biological, chemical, and physical properties of water.

Management. As opposed to business people who use this term to refer to the personnel in charge of the operation, wildlife managers use the term to imply the strategy and techniques that will be used to achieve specific goals.

Marsh. A community of water-tolerant, soft-bodied emergent plants and associated animals usually found in a basin of shallow water or on saturated soils fed by underground water sources.

Mitigation. Avoidance of or compensation for damages to natural habitats, resulting from human developments.

Mitigation bank. A system within the federally mandated mitigation process that allows land managers who may destroy wetlands at some future time to create, restore, or enhance wetlands in advance of the damage, and gain "points" that are "banked" for later use.

Model. A graphic or mathematical representation of a pattern or process in a biological system.

Niche. The role that an organism plays in an ecosystem: a primary producer such as a plant; a consumer such as a carnivore/animal eater (mink) or herbivore/plant eater (muskrat).

Nutrient. A chemical substance of value as a food component for plants and animals. Nutrients are incorporated by plants and then eaten by animals. Recycling of nutrients through the ecosystem is dependent on detritivores and decomposers.

Oligotrophic. As opposed to an enriched (eutrophic) lake or system, this refers to a less rich or even sterile body of water.

Omnivore. An animal that consumes both plant and animal material at the same time or during different seasons or at different stages in its life cycle.

Permafrost. Frozen water-rich soil and substrates in northern latitudes where thawing occurs only at the surface during midsummer.

pH. A scale to measure acidity (low index value) to alkalinity (higher numbers) based on the electrical characteristics of the constituents dissolved in the water. Wetlands often range from acid (pH 4.5) to quite alkaline (pH 8.5).

Phase (of marsh vegetation cycle). Because marshes often differ dramatically in vegetation as a result of water regimes, their vegetation patterns change and may be divided into phases such as drawdown, germination, emergent, submergent, and deep open-water.

Phosphorus. A chemical element essential as a component of living tissue and an essential nutrient that, in short supply, may limit plant productivity (hence, a common constituent of lawn and crop fertilizers).

Plankton. Small aquatic plants or animals that are suspended or may drift in the water. Plants such as algae (phytoplankton) may be responsible for significant levels of productivity in some open marshes, and the animals (zooplankton) include diverse microscopic protozoans and small crustaceans important as food for larger animals.

Pleistocene. A geological time period characterized by sheet glaciation in the northern United States, the most recent glaciation ranging from 8,000 years to 12,000 years ago. An important influence on landforms.

Population. Usually used in reference to a collection of individuals of one species making up the residents of a prescribed area, but may be used more broadly to describe mixed species.

Pothole. A term wildlife biologists use to describe the small, shallow ponds and marshes formed by Pleistocene glaciation in the grasslands of the northern United States and southern Canada. "Kettlehole" was the original term used.

Predator. An animal that pursues and eats living animals.

Producer. An organism such as a plant that produces organic matter by the natural process of nutrient and carbon dioxide intake in the presence of sunlight. Primary producers are plants that start with inorganic material and create organic tissue; secondary producers merely convert other organic tissue into animal tissue by means of food intake and growth.

Productivity. The end product of all producers, usually recorded as the number of gram calories produced per meter square per year.

Recharge zone. An area where water from wetlands moves downward into the underground water table.

Reservoir. An artificial body of water normally resulting from the impounding of water behind a dam. Small reservoirs also may be called impoundments.

Resource segregation. The division and use of a resource such as food by various species of animals. Reduced competition results and diversity of species probably is enhanced because they more effectively divide resources.

Restoration. The returning of a wetland (or other natural habitat) to its former state by modifying conditions responsible for the loss or change. Examples include replacing a natural dam and diverting water to recreate conditions suitable for seed germination and growth of emergent plants. Natural seed banks often remain even after many years, and a natural setting may develop with little help.

Restoration ecology. A new field of science dedicated to repairing damaged ecological communities.

Riverine. The channels, vegetated shallows, and bank associated with rivers, streams, and flowages.

Saline. Describes a body of water characterized by high levels of dissolved salts. Such bodies of water are common in arid regions as a result of evaporation, and along the coasts owing to the influence of ocean waters.

Seasonality. The influence of seasonal change on biological systems, such as plant growth, bird migration, and other events that are regulated by light, weather, and other seasonal events.

Section 404. A section of the Clean Water Act that gives authority to the U.S. Army Corps of Engineers to issue permits for dredging and filling in federally controlled waters.

Seed bank. A term applied by marsh ecologists and range scientists to the deposit of seeds in the soil that survive many years and that germinate when suitable conditions prevail. The durability of these seeds may make seeding unnecessary in marsh restoration and management.

Shredder. A term applied to animals of various sizes that break up vegetation into smaller parts through their feeding and cutting activities. They play an important role in the detrital cycle.

Siltation. The process whereby silt and other fine soil material accumulate in low areas or bodies of water such as marshes.

Social system. The sexual and familial organization evolved by a species or group that ensures reproduction and survival of offspring, and maintenance of a viable population.

Sociobiology. The study of the social aspects of interactions of animals that influence their adaptation and survival.

Spawning. The release of eggs by fish and aquatic invertebrates in specially selected sites that provide protection and suitable conditions for hatching.

Species. A population of similar and related organisms that reproduce but that are reproductively isolated from other similar groups.

Species association. A group of intermixed species found together. It does not imply organization or obligate relationship, but is merely a description of what is present at a particular time.

Species diversity. In general usage, the term implies the number of different species found in an area. In practice by ecologists, however, it tends to indicate a mathematical index (Species Diversity Index) that involves not only the number of species but relative numbers of each species as well.

Species richness. The number of species present, without reference to populations. Use of this term is preferable to species diversity to avoid confusion with the Species Diversity Index (see above).

Strata. Layers of vegetation of different heights above the ground. These layers create structural diversity (i.e., habitat diversity) and increase the habitats available to different species.

Succession. The process of change in plant and animal communities over time. The process can be very slow or fairly rapid. Wildlife managers commonly change habitat by manipulating succession, usually setting it back to "early" stages.

"Swimmer's itch." A rash produced by the larvae of various aquatic flukes that burrow into the human skin and die. Normally they seek intermediate hosts such as snails.

Threatened species. An uncommon to rare species given federal protection approaching that of an endangered species, but the designation is legally less restrictive on habitat protection or management.

Tide. The regular rise and fall of coastal waters produced by daily patterns of the moon's gravitational force in relation to the earth. Such changes may influence higher freshwater bodies as well, so both coastal salt water and freshwater may have tidal fluctuations.

Transpiration. The process whereby water is evaporated into the atmosphere from plant life processes. In a wetland, the amount of this water loss often exceeds that of direct evaporation of standing water.

Tundra. The vegetation zone in the far north or on high mountaintops characterized by low and non-woody vegetation that is adapted to conditions of cold, permafrost, and wind. Mosses, lichens, and sedges are prominent.

Turbidity. A measure of water clarity resulting from suspended material such as clay or algae. Measured with a secchi disk, a black-and-white-patterned or white disk lowered into the water to measure the depth to which it is visible.

Valuation. The process used by economists to place a value on a commodity. It is more difficult with items that are not for sale or that do not produce a product that can be sold or traded.

Vegetation structure. The physical attributes of vegetation (such as height, volume, configuration) that influence habitat appearance and diversity and attractiveness to animals.

Weed bank. An area that produces nuisance plants that may act as a source of seeds that contaminate other areas.

WET. An acronym for the U.S. Army Corps of Engineers' Wetland Evaluation Technique.

Wildlife surplus. The "surplus" in wild populations above those needed to maintain the species; regarded as harvestable by hunters.

References

Adamus, P. R., E. J. Clairain, Jr., R. D. Smith, and R. E. Young. 1987. Wetland evaluation technique (WET), Volume II: Methodology. Operational Draft Tech. Rep. Y-87. U.S. Army Corps of Engineers, Waterways Exp. Sta., Vicksburg, Mississippi.

Adamus, P. R., L. T. Stockwell, E. J. Clairain, Jr., M. E. Morrow, L. P. Rozas, and R. D. Smith. 1991. Wetland evaluation technique (WET), Volume I: Literature review and evaluation rationale. U.S. Army Corps of Engineers, Waterways Exp. Sta. Tech. Rep. WRP-DE-2. Nat. Tech. Info. Serv., Springfield, Virginia.

Addy, C. E., and L. G. MacNamara. 1948. Waterfowl management on small areas. Wildl. Mgmt. Inst. Washington, D.C. 80 pp.

Adomaitis, V. A., H. A. Kantrud, and J. A. Shoesmith. 1967. Some chemical characteristics of aeolian deposits of snow-soil on prairie wetlands. North Dakota Acad. Sci. 21:65–69.

Allred, E. R., P. W. Manson, G. M. Schwartz, P. Golany, and J. W. Reinke. 1971. Continuation of studies on the hydrology of ponds and small lakes, Univ. Minnesota Ag. Exp. Sta. Tech. Bull. 274. 62 pp.

Anderson, D. R., and F. A. Glover. 1967. Effects of water manipulation on waterfowl production and habitat. Trans. N. Am. Wildl. Nat. Resour. Conf. 32: 292-300.

Anderson, W. 1965. Waterfowl production in the vicinity of gull colonies. California Fish & Game 51:5–15.

Andrews, N. J., and D. C. Pratt. 1978. The potential of cattails (*Typha* spp.) as an energy source: productivity in managed stands. J. Minnesota Acad. Sci. 44:5-8.

Anthony, W. E. 1975. Basic economics of drainage. Univ. Minnesota Agric. Economist No. 568:3-5.

Atkinson, K. M. 1971. Further experiments in dispersal of phytoplankton by birds. Wildfowl Trust Ann. Rep. 22:98-99.

Atlantic Flyway Council. 1972. Techniques handbook of the waterfowl habitat development and management committee. 2nd ed. Atlantic Flyway Council, Boston, Massachusetts.

Barney, G. O. 1980. The global 2000 report to the President, Vol. I. U.S. Council on Environmental Quality and U.S. Dept. of State. 47 pp.

Bart, J., D. Allee, and M. Richmond. 1979. Using economics in defense of wildlife. Wildl. Soc. Bull. 7:139-144.

Beard, E. B. 1953. The importance of beaver in waterfowl management at Seney National Wildlife Refuge. J. Wildl. Mgmt. 17:398–436.

Beecher, W. J. 1942. Nesting birds and the vegetative substrate. Chicago Ornithological Society, Chicago, Illinois. 69 pp.

Bellrose, F. C. 1990. The history of wood duck management. Pp. 13–20 *in* Fredrickson, L. H. et al. (eds.), 1988 North American Wood Duck Symposium, Puxico, Missouri. 390 pp.

Bengtson, S. A. 1971. Variations in clutch-size in ducks in relation to the food supply. Ibis 113:523–526.

Bergman, R. D., R. L. Howard, K. F. Abraham, and M. W. Weller. 1977. Waterbirds and their wetland resources in relation to oil development at Storkersen Point, Alaska. U.S. Fish & Wildl. Serv. Resour. Publ. 129. 38 pp.

Bergman, R. D., P. Swain, and M. W. Weller. 1970. A comparative study of nesting Forster's and black terns. Wilson Bull. 82:435–444.

Beule, J. D. 1979. Control and management of cattails in southeastern Wisconsin wetlands. Wisconsin. Dept. Nat. Resour. Tech. Publ. No. 112. 39 pp.

Billington, C. 1938. The vegetation of the Cranbrook Lake bottom. Cranbrook Institute of Science, Bull. No. II. 24 pp.

Bishop, R. A., R. D. Andrews, and R. J. Bridges. 1979. Marsh management and its relationship to vegetation, waterfowl and muskrats. Proc. Iowa. Acad. Sci. 86:50-56.

Bishop, R. A., and R. Barratt. 1970. Use of artificial nest baskets by mallards. J. Wildl. Mgmt. 34:734-738.

Blair, C. L., and S. Sather-Blair. 1979. Highway planning coordination between resource and transportation agencies. South Dakota Coop. Wildl. Unit, Brookings. 148 pp.

Bossenmaier, E. F., and W. H. Marshall. 1958. Field-feeding by waterfowl in southwestern Manitoba. Wildl. Soc. Wildl. Monogr. 1. 32 pp.

Brenner, F. J., and J. J. Mondok. 1979. Waterfowl nesting rafts designed for fluctuating water levels. J. Wildl. Mgmt. 43:979-982.

Browder, J. A. 1978. A modeling study of water, wetlands, and wood storks. Pp. 325-346 *in* Sprunt, A. et al. (eds.), Wading birds. National Audubon Society Res. Rep. No. 7, New York. 381 pp.

Brown, M., and J. D. Dinsmore. 1986. Implications of marsh size and isolation for marsh management. J. Wildl. Mgmt. 50:392–397.

Brown, M., and J. J. Dinsmore. 1991. Area-dependent changes in bird densities in Iowa marshes. J. Iowa Acad. Sci. 98(3):124–126.

Buchsbaum, R., and M. Buchsbaum. 1957. Basic ecology. Boxwood Press, Pittsburgh, Pennsylvania. 192 pp.

Buech, R. R. 1985. Beaver in water impoundments: understanding a problem of water-level management. Pp. 95–105 *in* Knighton, M. D. (ed.), Proc. of water impoundments for wildlife: a habitat management workshop, 1982. U.S. For. Serv. N. Cent. Forest Exp. Sta. St. Paul, Minnesota. 136 pp.

Cahn, A. R. 1929. The effect of carp on a small lake; the carp as a dominant. Ecol. 10:271-274.

Carignan, R., and J. Kalff. 1980. Phosphorous sources for aquatic weeds: water or sediments? Science 207:987-989.

Cartwright, B. W. 1942. Regulated burning as a marsh management technique. Trans. N. Am. Wildl. Conf. 7:257-263.

Cartwright, B. W. 1946. Muskrats, duck production and marsh management. Trans. N. Am. Wildl. Conf. 11:454–457.

Catchpole, C. K., and C. F. Tydeman. 1975. Gravel pits as new wetland habitats for the conservation of breeding bird communities. Biol. Conser. 8:47-59.

Chabreck, R. H. 1976. Management of wetlands for wildlife habitat improvement. Pp. 226–233 *in* Wiley, M. (ed.), Estuarine processes, Vol. 1. Academic Press, New York.

Chabreck, R. H. 1979. Wildlife harvest in wetlands of the United States. Pp. 618-631 *in* Greeson, P. E., J. R. Clark, and J. E. Clark (eds.), Wetland functions and values: the state of our understanding. Amer. Water Resour. Assoc. Minneapolis, Minnesota. 674 pp.

Chabreck, R. H. 1988. Coastal marshes; ecology and wildlife management. Univ. Minnesota Press, Minneapolis. 138 pp.

Chabreck, R. H., J. E. Holcombe, R. G. Linscombe, and N. E. Kinler. 1982. Winter foods of river otters from saline and fresh environments in Louisiana. Proc. Ann. Conf. S.E. Assoc. Fish & Wildl. Agencies 36:1-20.

Choate, J. S. 1972. Effects of stream channeling on wetlands in a Minnesota watershed. J. Wildl. Mgmt. 36:940-944.

Christiansen, J. E., and J. B. Low. 1970. Water requirements of waterfowl marshlands in northern Utah. Utah Div. of Fish & Game. No. 69-12. 108 pp.

Clements, F. E. 1916. Plant succession: an analysis of the development of vegetation. Public. 242, Carnegie Institute, Washington, D.C. 512 pp.

Cook, A., and C. F. Powers. 1958. Early biochemical changes in the soils and waters of artificially created marshes in New York. N.Y. Fish & Game J. 5:9-65.

Cowardin, L. M., V. Carter, F. C. Golet, and E. T. LaRoe. 1979. Classification of wetlands and deepwater habitats of the United States. U.S. Fish & Wildl. Serv. Off. of Biol. Serv. 103 pp.

Cowardin, L. M., and D. H. Johnson. 1973. A preliminary classification of wetland plant communities in north-central Minnesota. U. S. Fish & Wildl. Serv., Spec. Sci. Rep. Wildl. 168. 33 pp.

Crail, L. R. 1955. Viability of smartweed and millet seeds in relation to marsh management in Missouri. Missouri Conserv. Comm. PR Project Rep. 13-R-5. 16 pp.

Crocker, W. 1938. Life span of seeds. Bot. Rev. 4:235-272.

Cuthbert, N. L. 1954. A nesting study of the black tern in Michigan. Auk 71: 36–63.

Dahl, T. E. 1990, Wetland losses in the United States 1780's to 1980's. U.S. Dept. of Int., Fish & Wildl. Serv., Washington, D.C. 21 pp.

Danell, K., and K. Sjoberg. 1977. Seasonal emergence of chironomids in relation to egglaying and hatching of ducks in a restored lake. Wildfowl 28:129-135.

de la Cruz, A. A. 1979. Production and transport of detritus in wetlands. Pp. 162–174 *in* Greeson, P. E., J. R. Clark, and J. E. Clark (eds.), Wetland functions and values: the state of our understanding. Amer. Water Resour. Assoc. Minneapolis, Minnesota. 674 pp.

DeVlaming, V., and V. W. Proctor. 1968. Dispersal of aquatic organisms: viability of seeds recovered from the droppings of captive killdeer and mallard ducks. Am. J. Bot. 55:20–26.

Dozier, H. L. 1947. Salinity as a factor in Atlantic Coast tidewater muskrat production. Trans. N. Am. Wild. Conf. 12:398–420.

Duebbert, H. E. 1969. The ecology of Malheur Lake. U.S. Fish & Wildl. Serv. Refuge Leaf. No. 12. 24 pp.

Duebbert, H. F., and J. T. Lokemoen. 1976. Duck nesting in fields of undisturbed grass and legume cover. J. Wildl. Mgmt. 40:39–49.

Eisenlohr, W. S., Jr., C. E. Sloan, and J. S. Shjeflo. 1972. Hydrologic investigations of prairie potholes in North Dakota, 1959-1968. Geol. Surv. Prof. Paper 585-A. 102 pp.

Elton, C. S. 1958. The ecology of invasions of animals and plants. Methuen, London. 181 pp.

Emlen, S. T., and H. W. Ambrose, III. 1970. Feeding interactions of snowy egrets and red-breasted mergansers. Auk 87:164-165.

Erickson, R. E., R. L. Linder, and K. W. Harmon. 1979. Stream channelization (P.L. 33-556) increased wetland losses in the Dakotas. Wildl. Soc. Bull. 7:71-78.

Eriksson, M. O. G. 1978. Lake selection by goldeneye ducklings in relation to the abundance of food. Wildfowl 29:81-85.

Errington, P. L. 1937. Drowning as a cause of mortality in muskrats. J. Mamm. 18:497-500.

Errington, P. L. 1957. Of men and marshes. Macmillan Co., New York. 150 pp.

Errington, P. L. 1963. Muskrat populations. Iowa St. Univ. Press, Ames. 665 pp.

Errington, P. L., and T. S. Scott. 1945. Reduction in productivity of muskrat pelts on an Iowa marsh through depredations of red foxes. J. Agric. Res. 71: 137–148.

Errington, P. L., R. J. Siglin, and R. C. Clark. 1963. The decline of a muskrat population. J. Wildl. Mgmt. 27:1–8.

Etherington, J. R. 1983. Wetland ecology. Studies in Biology, No. 154. Edward Arnold, London. 66 pp.

Evans, C. D., and K. E. Black. 1956. Duck production studies on the prairie potholes of South Dakota. U.S. Fish & Wildl. Serv. Spec. Sci. Rep. (Wildl.) No. 32. 59 pp.

Ewel, K. C., and H. T. Odum. 1984. Cypress swamps. Florida Presses of Florida, Gainesville. 472 pp.

Faaborg, J. 1976. Habitat selection and territorial behavior of the small grebes of North Dakota. Wilson Bull. 88:390–399.

Fassett, N. C. 1940. A manual of aquatic plants. McGraw-Hill Book Co., New York, 382 pp.

Federal Interagency Committee for Wetland Delineation. 1989.

Federal manual for identifying and delineating jurisdictional wetlands. U.S. Army Corps of Engineers, U.S. Environmental Protection Agency, U.S. Fish & Wildlife Service, and U.S.D.A. Soil Conservation Service Cooperative technical publication, Washington, D.C. 76 pp. plus appendices.

Flake, L. D. 1979. Wetland diversity and waterfowl. Pp. 312-319 *in* Greeson, P. E., J. R. Clark, and J. E. Clark (eds.), Wetland functions and values: the state of our understanding. Amer. Water Resour. Assoc., Minneapolis, Minnesota. 674 pp.

Fog, J. 1980. Methods and results of wetland management for waterfowl. Acta Ornithologica XVII (12):147-160.

Forney, J. L. 1968. Production of young northern pike in a regulated marsh. N.Y. Fish & Game 15:143-154.

Foster, J. H. 1979. Measuring the social value of wetland benefits. Pp. 84-92 *in* Greeson, P. E., J. R. Clark, and J. E. Clark (eds.), Wetlands functions and values: the state of our understanding. Amer. Water Resour. Assoc., Minneapolis, Minnesota. 674 pp.

Frayer, W. E., T. J. Monahan, D. C. Bowden, and F. A. Graybill. 1983. Status and trends of wetlands and deepwater habitats in the conterminous United States, 1950's to 1970's. Dept. of Forest and Wood Sciences, Colorado St. Univ., Fort Collins. 32 pp.

Fredrickson, L. H., and D. L. Batema. 1992. Greentree reservoir management handbook. Gaylord Memorial Lab., Univ. Missouri, Puxico. 88 pp.

Fredrickson, L. H., and T. Scott Taylor. 1982. Management of seasonally flooded impoundments for wildlife. U.S. Fish & Wildl. Serv. Resour. Publ. 148, Washington, D.C. 29 pp.

Fritzell, P. A. 1979. American wetlands as cultural symbol: place of wetlands in American culture. Pp. 523-534 *in* Greeson, P. E., J. R. Clark, and J. E. Clark (eds.), Wetland functions and values: the state of our understanding. Amer. Water Resour. Assoc., Minneapolis, Minnesota. 674 pp.

Gale, W. F. 1975. Bottom fauna of a segment of Pool 19, Mississippi River, near Fort Madison, Iowa, 1967-1968. Iowa St. J. Res. 49:353-372.

Ganning, B., and F. Wulff. 1969. The effects of bird droppings on chemical and biological dynamics in brackish water rockpools. Oikos 20:274-289.

Gasaway, R. D., and T. F. Drda. 1977. Effects of grass carp introduction on waterfowl habitat. Trans. N. Am. Wildl. Nat. Resour. Conf. 42:73-85.

Geis, J. W. 1979. Shoreline processes affecting the distribution of wetland habitat. Trans. N. Am. Wildl. Nat. Resour. Conf. 44:529-542.

Giles, R. H., Jr. (ed.). 1969. Wildlife Management Techniques, 3rd ed., revised. The Wildlife Society, Washington, D.C. 623 pp.

Glazener, W. C. 1946. Food habits of wild geese on the Gulf Coast of Texas. J. Wildl. Mgmt. 10:322-329.

Gleason, H. A. 1917. The structure and development of the plant association. Bull. Torrey Bot. Club 44:463-481.

Golet, F. C., and J. S. Larson. 1974. Classification of freshwater wetlands in the glaciated northeast. U. S. Fish & Wildl. Serv. Resour. Publ. 116. 56 pp.

Goss, W. L. 1924. The viability of buried seeds. J. Agric. Res. 29:349-362.

Gosselink, J. C. 1984. The ecology of delta marshes of coastal Louisiana: a community profile. U.S. Fish & Wildl. Serv. FWS/OBS-84/09. 134 pp.

Gosselink, J. C., E. P. Odum, and R. M. Pope. 1973. The value of the tidal marsh. Public. No. LSU-SG-74-03. Center for Wetland Resources, Baton Rouge, Louisiana. 30 pp.

Grace, J. B. 1989. Effects of water depth on *Typha latifolia* and *Typha domingensis*. Amer. J. Bot. 76:762-768.

Gupta, R. R., and J. H. Foster. 1975. Economics criteria for freshwater wetland policy in Massachusetts. Am. J. Agric. Econ. 57:40-45.

Hammack, J., and G. M. Brown, Jr. 1974. Waterfowl and wetlands: toward bioeconomic analysis. Johns Hopkins Univ. Press, Baltimore. 95 pp.

Hammer, D. A.(ed.). 1989. Constructed wetlands for wastewater treatment. Lewis Publishers, Inc., Chelsea, Michigan. 831 pp.

Hammer, D. A. 1992. Creating freshwater wetlands. Lewis Publishers, Inc., Chelsea, Michigan. 298 pp.

Hands, H. M., M. R. Ryan, and J. W. Smith. 1991. Migrant shorebird use of marsh, moist-soil, and flooded agricultural habitats. Wildl. Soc. Bull. 19: 457-464.

Harmon, K. W. 1979. Mitigating losses of private wetlands: the North Dakota situation. Pp. 157-163 *in* Swanson, G. A. (ed.), The mitigation symposium. Gen. Tech. Rept. RM-65. U.S. For. Serv. Rocky Mtn. For. & Range Exp. Sta., Fort Collins, Colorado. 684 pp.

Harris, S. W., and W. H. Marshall. 1963. Ecology of water-level manipulations of a northern marsh. Ecol. 44:331-342.

Harrison, J. 1970. Creating a wetland habitat. Bird Study 17:111-122.

Harter, R. D. 1966. The effect of water levels on soil chemistry and plant growth of the Magee Marsh Wildlife Area. Ohio Dept. Nat. Res. Game Monogr. No. 2. 36 pp.

Have, M. R. 1973. Effects of migratory waterfowl on water quality at the Montezuma National Wildlife Refuge, Seneca County, New York. Res. U.S. Geol. Surv. 1:725-734.

Heding, R. 1964. Game fish nurseries. Wisconsin Conserv. Bull. 29:7.

Hochbaum, H. A. 1944. The canvasback on a prairie marsh. Amer. Wildl. Inst., Washington, D.C.

Hoffpauir, C. M. 1968. Burning for coastal marsh management. Pp. 134-139 *in* Newson, J. (ed.), Proc. of the marsh and estuarine management symposium. Baton Rouge, Louisiana.

Hohman, W. L. 1977. Invertebrate habitat preferences in several contiguous Minnesota wetlands. M.S. thesis, Univ. North Dakota, Fargo. 79 pp.

Horwitz, E. L. 1978. Our nation's wetlands—an interagency task force report. GPO, Washington, D.C. 70 pp.

Hotchkiss, N. 1970. Common marsh plants of the United States and Canada. U.S. Fish & Wildl. Serv. Resour. Publ. 93. Washington, D.C. 99 pp.

Jeglum, J. K., A. N. Boissonneau, and V. F. Haavisto. 1974. Toward a wetland classification for Ontario. Can. For. Serv., Sault Ste. Marie, Ont. Inf. Rep. O-X-215. 54 pp.

Joyner, D. E. 1980. Influence of invertebrates on pond selection by ducks in Ontario. J. Wildl. Mgmt. 44:700-705.

Kadlec, J. A. 1962. Effects of a drawdown on a waterfowl impoundment. Ecol. 43:267-281.

Kadlec, J. A. 1979. Nitrogen and phosphorus dynamics in inland freshwater wetlands. Pp. 17-41 *in* Bookout, T. A. (ed.), Waterfowl and wetlands—an integrated review. Proc. of a symposium of the 39th Fish & Wildl. Conf. Madison, Wisconsin. 1977.

Kadlec, R. H., and J. A. Kadlec. 1979. Wetlands and water quality. Pp. 436-456 *in* Greeson, P. E., J. R. Clark, and J. E. Clark (eds.), Wetland functions and values: the state of our understanding. Amer. Water Resour. Assoc., Minneapolis, Minnesota. 674 pp.

Kaminski, R. M., and H. H. Prince. 1981. Dabbling duck activity and foraging response to aquatic macroinvertebrates. Auk 98:115-126.

Kaminski, R. M., and H. H. Prince. 1984. Dabbling duck—habitat associations during spring in the Delta Marsh, Manitoba. J. Wildl. Mgmt. 48:37-50.

Kantrud, H. 1986. Effects of vegetation manipulation on breeding waterfowl in prairie wetland—a literature review. U.S. Fish & Wildl. Serv. Tech. Rep. 3. 15 pp.

Kantrud, H., G. L. Krapu, and G. A. Swanson. 1989. Prairie basin wetlands of the Dakotas: a community profile. U.S. Fish & Wildl. Serv. Biol. Rep. 85(7.28). 116 pp.

Keith, L. B. 1964. Some social and economic values of the recreational use of Horicon Marsh, Wisconsin. Univ. Wisconsin Res. Bull. 246. 16 pp.

Kellert, S. R. 1979. Public attitudes toward critical wildlife and natural habitat issues, phase I. Nat. Tech. Info. Serv., PB 80–138332. U.S. Dept. Commerce, Washington, D.C. 138 pp.

Kelting, R. W. and W. T. Penfound. 1950. The vegetation of stock pond dams in central Oklahoma. Am. Midl. Nat. 44:69-75.

Kendeigh, S. C. 1961. Animal ecology. Prentice-Hall, Inc., Englewood Cliffs, New Jersey. 468 pp.

Kiel, W. H., Jr., A. S. Hawkins, and N. G. Perret. 1972. Waterfowl habitat trends in the aspen parkland of Manitoba. Can. Wildl. Serv. Rep. Ser. 18. 61 pp.

King, D. R., and G. S. Hunt. 1967. Effect of carp on vegetation in a Lake Erie marsh. J. Wildl. Mgmt. 31:181-188.

Kirsch, L. M. 1969. Waterfowl production in relation to grazing. J. Wildl. Mgmt. 33:821-828.

Kleinert, S. J. 1970. Production of northern pike in a managed marsh, Lake Ripley, Wisconsin. Wisconsin Dept. Nat. Res. Rep. 49. 19 pp.

Koskimies, J. 1957. Terns and gulls as features of habitat recognition for birds nesting in their colonies. Ornis Fennica 34:1-6.

Krapu, G. L. 1974. Feeding ecology of pintail hens during reproduction. Auk 91: 278-290.

Krapu, G. L., and H. F. Duebbert. 1974. A biological survey of Kraft Slough. Prairie Naturalist 6:33-55.

Krapu, G. L., D. R. Parsons, and M. W. Weller. 1970. Waterfowl in relation to land use and water levels on the Spring Run area. Iowa St. J. Sci. 44:437-452.

Krapu, G. L. and K. J. Reinecke. 1992. Foraging ecology and nutrition. Pages 1–29 in Batt, B. D. J. et al. (eds.), Ecology and management of breeding waterfowl. Univ. Minnesota Press, Minneapolis. 635 pp.

Krecker, F. H. 1939. A comparative study of the animal population of certain submerged aquatic plants. Ecol. 20:553-562.

Krummes, W. T. 1941. The muskrat: a factor in waterfowl habitat management. Trans. N. Am. Wildl. Conf. 5:395-398.

Kushlan, J. A. 1974. Quantitative sampling of fish populations in shallow, freshwater environments. Trans. Am. Fish. Soc. 103:348-352.

Kusler, J. A. 1983. Our national wetland heritage. Environmental Law Institute, Washington, D.C. 167 pp.

Kusler, J. A., and M. E. Kentula (eds.). 1990. Wetland creation and restoration: the status of the science. Island Press, Covelo, California.

Lagler, K. F. 1956. The pike, Esox lucius Linnaeus, in relation to waterfowl on the Seney National Wildlife Refuge, Michigan. J. Wildl. Mgmt. 20:114-124.

Larson, J. S. 1971. Progress toward a decision-making model for public management of fresh-water wetlands. Trans. N. Am. Wildl. Nat. Resour. Conf. 36:376-382.

Larson, J. S. 1975. Evaluation models for public management of freshwater wetlands. Trans. N. Am. Wildl. Nat. Resour. Conf. 40:220-228.

Laycock, G. 1965. The sign of the flying goose. Natural History Press, Garden City, New York. 299 pp.

Leschisin, D. A., G. L. Williams, and M. W. Weller. 1992. Factors affecting waterfowl use of constructed wetlands in northwestern Minnesota. Wetlands 12:178–183.

Leitch, J. A., and L. E. Danielson. 1979. Social, economic and institutional incentives to drain or preserve prairie wetlands. Univ. Minnesota Dept. Agric. & Appl. Econ., Rep. EC79-6. 78 pp.

Leitch, W. G., and R. M. Kaminski. 1985. Long-term wetland-waterfowl trends in Saskatchewan grassland. J. Wildl. Mgmt. 49:212–222.

Linde, A. F. 1969. Techniques for wetlands management. Wisconsin Dept. Nat. Resour. Rep. 45. 156 pp.

Linde, A. F. 1985. Vegetation management in water impoundments: alternative and supplements to water-level control. Pp. 51–60 *in* Knighton, M.D. (ed.), Proc. of water impoundments for wildlife: a habitat management workshop. U.S. For. Serv. N. Cent. For. Exp. Sta. St. Paul, Minnesota. 136 pp.

Linde, A. F., T. Janisch, and D. Smith. 1976. Cattail—the significance of its growth, phenology, and carbohydrate storage to its control and management. Wisconsin Dept. Nat. Resour. Tech. Bull. 94. 27 pp.

Linduska, J. P. (ed.). 1964. Waterfowl tomorrow. U.S. Fish & Wildl. Serv., Washington, D.C. 770 pp.

Lingle, G. R., and N. F. Sloan. 1980. Food habits of white pelicans during 1976 and 1977 at Chase Lake National Wildlife Refuge, North Dakota. Wilson Bull. 92:123-125.

Livingston, R. J., and O. L. Loucks. 1979. Productivity, trophic interactions, and food web relationships in wetlands and associated systems. Pp. 101–119 *in* Greeson, P. E., J. R. Clark, and J. E. Clark (eds.), Wetland functions and values: the state of our understanding. Amer. Water Resour. Assoc., Minneapolis, Minnesota. 674 pp.

Lokemoen, J. T., and R. O. Woodward. 1992. Nesting waterfowl and water birds on natural islands in the Dakotas and Montana. Wildl. Soc. Bull. 20:163–171.

Lueschow, L. A. 1972. Biology and control of aquatic nuisances in recreational waters. Wisconsin Dept. Nat. Resour. Tech. Bull. 57. 35 pp.

Lynch, J. J. 1941. The place of burning in the management of Gulf Coast wildlife refuges. J. Wildl. Mgmt. 5:454-459.

Lynch, J. J., T. O. O'Neil, and D. W. Lay. 1947. Management significance of damage by geese and muskrats to Gulf Coast marshes. J. Wildl. Mgmt. 11: 50–76.

MacBride, T. H. 1909. The geology of Hamilton and Wright Counties. Iowa. Geol. Surv. 20:101-138.

Mack, G. D., and L. D. Flake. 1980. Habitat relationships of waterfowl broods in South Dakota stock ponds. J. Wildl. Mgmt. 44:695-700.

Madson, R. 1980. Prairie wetlands; a resource threatened. N. Midwest Reg. Audubon Newsletter 2:1-2.

Malone, C. R. 1965. Dispersal of aquatic gastropods via the intestinal tract of water birds. Nautilus 78:135-139.

Manny, B. A., R. G. Wetzel, and W. C. Johnson. 1975. Annual contribution of carbon, nitrogen, and phosphorus by migrant Canada geese to a hardwater lake. Verh. Internat. Verein. Limnol. 19:945-951.

Martin, A. C., R. C. Erickson, and J. H. Steenis. 1957. Improving duck marshes by weed control. U.S. Fish & Wildl. Serv. Circ. 19-rev. Washington, D.C. 60 pp.

Martin, A. C., N. Hotchkiss, F. M. Uhler, and W. S. Bourn. 1953. Classification of wetlands of the United States. U.S. Fish & Wildl. Serv., Spec. Sci. Rep. Wildl. 20. 14 pp.

Martin, E. M., A. N. Novara, P. H. Geissler, and S. M. Carney. 1989. Preliminary estimates of waterfowl harvest and hunter activity in the United States during the 1988 hunting season. U.S. Fish & Wildl. Serv. Off. Migratory Bird Mgmt., Laurel, Maryland. 23 pp.

Mathiak, H. 1965. Pothole blasting for wildlife. Publ. 352. Wisconsin Cons. Dept. Madison. 31 pp.

Mathiak, H. A., and A. F. Linde. 1956. Studies on level ditching for marsh management. Wisconsin Con. Dept. Tech. Wildl. Bull. 12. 48 pp.

Mathisen, J. E. 1985. Wildlife impoundments in the north central States: why do we need them? Pp.

23–30 *in* Knighton, M. D. (ed.), Proc. of water impoundments for wildlife: a habitat management workshop. U.S. For. Serv. N. Cent. For. Exp. Sta. St. Paul, Minnesota. 136 pp.

Maxwell, G. 1957. People of the reeds. Pyramid Publications, New York. 222 pp.

McAndrews, J. H., R. E. Stewart, Jr., and R. C. Bright. 1967. Paleoecology of a prairie pothole; a preliminary report. Pp. 101–113 *in* Clayton, Lee, and Freers (eds.), Midwestern Friends of the Pleistocene guidebook, 18th Ann. Field Conf. North Dakota Geol. Surv. Misc. Ser. 30.

McKnight, S. K. 1992. Transplanted seed bank response to drawdown time in a created wetland in east Texas. Wetlands 12(2):91–98.

Meanley, B. 1971. Blackbirds and the southern rice crop. U.S. Fish & Wildl. Serv. Res. Publ. 100, Washington, D.C. 64 pp.

Meeks, R. L. 1969. The effect of drawdown date on wetland plant succession. J. Wildl. Mgmt. 33:817–821.

Millar, J. B. 1971. Shoreline-area as a factor in rate of water loss from small sloughs. J. Hydrol. 14:259–284.

Millar, J. B. 1976. Wetland classification in western Canada: a guide to marshes and shallow open water wetlands in the grasslands and parklands of the Prairie Provinces. Can. Wildl. Serv. Rep. Ser. 37. 38 pp.

Miller, R. B., and R. C. Thomas. 1956. Alberta's pothole trout fisheries. Tran. Am. Fish. Soc. 86:261–268.

Moore, I. D., and C. L. Larson. 1979. Effects of drainage projects on surface runoff from small depressional watersheds in the North Central Region. Univ. Minnesota Water Resour. Res. Center Bull. 99. 225 pp.

Momot, W. T., and H. Gowing. 1978. The dynamics of crayfish and their role in ecosystems. Am. Midl. Nat. 99:10–35.

National Wetlands Policy Forum. 1988. Protecting America's wetlands: an action agenda. The Conservation Foundation, Washington, D. C. 69 pp.

Needham, J. G., and P. R. Needham. 1941. A guide to the study of freshwater biology. Comstock Publ., Co., Ithaca, New York. 89 pp.

Neely, R. K., and J. L. Baker. 1989. Nitrogen and phosphorous dynamics and the fate of agricultural runoff. Pp. 92–131 *in* van der Valk, A. G. (ed.), Northern prairie wetlands. Iowa St. Univ. Press, Ames. 400 pp.

Nelson, N. F., and R. H. Dietz. 1966. Cattail control methods in Utah. Utah St. Dept. of Fish & Game Publ. No. 66-2. 31 pp.

Nickell, W. P. 1966. Common terns nest on muskrat lodges and floating cattail mats. Wilson Bull. 78:123–124.

Odum, E. P. 1971. Fundamentals of ecology. W. B. Saunders Co., Philadelphia, Pennsylvania. 574 pp.

Orians, G. H. 1961. The ecology of blackbird (*Agelaius*) social systems. Ecol. Monogr. 31:285–312.

Orians, G. H., and H. S. Horn. 1969. Overlap in foods of four species of blackbirds in the potholes of central Washington. Ecol. 50:930–938.

Ortego, B., R. B. Hamilton, and R. E. Noble. 1976. Bird usage by habitat types in a large freshwater lake. Proc. S.E. Fish & Game Conf. 13:627–633.

Osvald, S., and C. W. Belin. 1979. Corps permit processing. Pp. 50–56 *in* Greeson, P. E., J. R. Clark, and J. E. Clark (eds.), Wetland functions and values: the state of our understanding. Amer. Water Resour. Assoc., Minneapolis, Minnesota. 674 pp.

Ovington, J. D., and W. H. Pearsall. 1956. Production ecology II, shoot production in phragmites in relation to habitat. Oikos 7:206–214.

Palmisano, A. W. 1973. Habitat preferences of waterfowl and fur animals in the northern Gulf Coast marshes. Pp. 163–190 *in* Chabreck, R. H. (ed.), Proc. 2nd coastal marsh and estuary management symposium. Louisiana. St. Univ., Baton Rouge.

Paterson, C. G., and C. H. Fernando. 1971. A comparison of a simple corer and an Ekman grab for sampling shallow-water benthos. J. Fish. Res. Bd. Canada 28:365-368.

Patterson, J. H. 1976. The role of environmental heterogeneity in the regulation of duck populations. J. Wildl. Mgmt. 40:22-32.

Payne, N. F. 1992. Techniques for wildlife habitat management of wetlands. McGraw-Hill, New York. 566 pp.

Pederson, R. L., and A. G. van der Valk. 1984. Vegetation change and seed banks in marshes: ecological and management implications. Trans. N. Am. Wildl. Nat. Resour. Conf. 49:271-280.

Pennak, R. W. 1978. Fresh-water invertebrates of the United States. 2nd ed. John Wiley, New York. 803 pp.

Peterka, J. J. 1989. Fishes in northern prairie wetlands. Pp. 302–315 *in* van der Valk, A. G. (ed.), Northern prairie wetlands. Iowa St. Univ. Press, Ames. 400 pp.

Priegel, G. R. 1970. Reproduction and early life history of the walleye in the Lake Winnebago region. Wisconsin. Dept. Nat. Resour. Tech. Bull. 45. 105 pp.

Provost, M. W. 1947. Nesting birds in the marshes of northwest Iowa. Am. Midl. Nat. 38:485-503.

Provost, M. W. 1948. Marsh-blasting as a wildlife management technique. J. Wildl. Mgmt. 12:350–387.

Quade, H. W. 1969. Cladoceran faunas associated with aquatic macrophytes in some lakes in northwestern Minnesota. Ecol. 50:170-179.

Ranwell, D. S. 1967. Introduced aquatic, fresh-water and salt marsh-case histories and ecological effects. Proc. & Papers IUCN Tech. Mtg. 10, IUCN Publ. New Ser. No. 9:27-37.

Reed, P. B., Jr. 1988. National list of plant species that occur in wetlands: national summary. U.S. Fish & Wildl. Serv. Biol. Rep. 88(24). GPO, Washington, D.C. 244 pp.

Reichholf, J. 1976. The possible use of the aquatic bird communities as indicators for the ecological conditions of wetlands. Landschaft Stadt. 8:125-129.

Reid, G. K. 1961. Ecology of inland waters and estuaries. D. Van Nostrand Co., New York. 375 pp.

Riemer, D. N. 1984. Introduction to freshwater vegetation. AVI Publ. Co., Inc., Westport, Connecticut. 207 pp.

Robel, R. J. 1961. Water depth and turbidity in relation to growth of sago pondweed. J. Wildl. Mgmt. 25:436–438.

Robel, R. J. 1962. Changes in submersed vegetation following a change in water level. J. Wildl. Mgmt. 26:221-224.

Rogers, J. P. 1959. Low water and lesser scaup reproduction near Erickson, Manitoba. Trans. N. Am. Wildl. Conf. 24:216-224.

Rogers, J. P., J. D. Nichols, F. W. Martin, and C. F. Kimball. 1979. An examination of harvest and survival rates of ducks in relation to hunting. Trans. N. Am. Wildl. Nat. Resour. Conf. 44:114-126.

Rosenbaum, N. 1979. Enforcing wetlands regulations. Pp. 43-49 *in* Greeson, P. E., J. R. Clark, and J. E. Clark (eds.), Wetland functions and values: the state of our understanding. Amer. Water Resour. Assoc., Minneapolis, Minnesota, 674 pp.

Rosine, W. N. 1955. The distribution of invertebrates on submerged aquatic plant surfaces in Muskee Lake, Colorado. Ecol. 36:308-314.

Rundel, W. D., and L. H. Fredrickson. 1981. Managing seasonally flooded impoundments for migrant rails and shorebirds. Wildl. Soc. Bull. 9:80-87.

Ryther, J. H., T. A. DeBusk, M. D. Hanisak, and L. D. Williams. 1979. Fresh water macrophytes for energy and waste water treatment. Pp. 652-660 *in* Greeson, P. E., J. R. Clark, and J. E. Clark (eds.), Wetland functions and values: the state of our understanding. Amer. Water Resour. Assoc., Minneapolis, Minnesota. 674 pp.

Sanderson, G. C. 1974. Habitat—key to wildlife perpetuation; aquatic areas. Pp. 21–40 *in* How do we achieve and maintain variety and optimum numbers of wildlife? Symposium, Nat. Wildl. Fed., Washington, D.C. 103 pp.

Schroeder, L. D. 1973. A literature review on the role of invertebrates in waterfowl management. Colorado Div. Wildl. Spec. Rep. No. 29. 13 pp.

Sculthorpe, C. D. 1967. The biology of aquatic vascular plants. Edward Arnold, Ltd., London. 610 pp.

Sealy, S. G. 1978. Clutch size and nest placement of the pied-billed grebe in Manitoba. Wilson Bull. 90:301-302.

Sewell, R. W., and K. F. Higgins. 1991. Pp. 108–133 *in* Webb, F. J., Jr., (ed.), Proc. Ann. Conf. of Wetland Restoration and Creation, 18. Hillsborough Comm. Coll., Tampa, Florida.

Shaw, S. P., and C. G. Fredine. 1956. Wetlands of the United States. U.S. Fish & Wildl. Serv. Circ. 39. Washington, D.C. 67 pp.

Shearer, L. A., B. J. Jahn, and L. Ienz. 1969. Deterioration of duck foods when flooded. J. Wildl. Mgmt. 33:1012-1015.

Shindler, D. W. 1974. Eutrophication and recovery in experimental lakes: implications for lake management. Science 184:897-899.

Shull, G. H. 1914. The longevity of submerged seed. Plant World 17:329-337.

Singleton, J. R. 1951. Production and utilization of waterfowl food plants on the east Texas coast. J. Wildl. Mgmt. 15:46–56.

Sloan, C. E. 1972. Ground-water hydrology of prairie potholes in North Dakota. U.S. Geol. Surv. Prof. Paper 585-C. 28 pp.

Smith, L. M. 1985a. Fire and herbivory in a Great Salt Lake marsh. Ecol. 66:259-265.

Smith, L. M. 1985b. Predictions of vegetation change following fire in a Great Salt Lake marsh. Aquatic Bot. 21:43-51.

Smith, R. H. 1964. Experimental control of purple loosestrife (Lythrum salicaria). N.Y. Fish & Game J. 11:35–46.

Smith, R. I. 1970. Response of pintail breeding populations to drought. J. Wildl. Mgmt. 34:943-946.

Smith, R. L. 1980. Ecology and field biology, 3rd ed. Harper and Row, Publishers, New York. 835 pp.

Stearns, L. A., D. MacCreasy, and F. C. Daigh. 1940. Effect of ditching for mosquito control on the muskrats of a Delaware tidal marsh. Univ. Delaware Agric. Exp. Sta. Bull. No. 225. 55 pp.

Stewart, R. E., and H. A. Kantrud. 1971. Classification of natural ponds and lakes in the glaciated prairie region. U.S. Fish & Wildl. Serv. Resour. Publ. 92. 57 pp.

Stewart, R. E., and H. A. Kantrud. 1972. Vegetation of prairie potholes, North Dakota, in relation to quality of water and other environmental factors. U.S. Geol. Surv. Prof. Paper 585-D. 36 pp.

Stewart, R. E., and H. A. Kantrud. 1973. Ecological distribution of breeding waterfowl populations in North Dakota. J. Wildl. Mgmt. 37:39-50.

Stone, C. P., and D. F. Mott. 1973. Bird damage to ripening field corn in the United States, 1971. U.S. Fish & Wildl. Serv. Wildl. Leaf. 505:1-8. Washington, D.C.

Strohmeyer, D. L., and L. H. Fredrickson. 1967. An evaluation of dynamited potholes in northwest Iowa. J. Wildl. Mgmt. 31:525-532.

Sugden, L. G. 1978. Canvasback habitat use and production in Saskatchewan parklands. Can. Wildl. Serv. Occas. Paper No. 34. 30 pp.

Sugden, L. G., and D. A. Benson. 1970. An evaluation of loafing rafts for attracting ducks. J. Wildl. Mgmt. 34:340-343.

Svedarsky, D., and R. D. Crawford (eds.). 1982. Wildlife values of gravel pits. Miscellaneous Publ. 17-1982, Univ. Minnesota Agric. Exp. Sta., St. Paul, Minnesota. 249 pp.

Swanson, G. A., and M. I. Meyer. 1973. The role of invertebrates in the feeding ecology of Anatids during the breeding season. Pp. 143-185 *in* The waterfowl habitat management Symposium. Moncton, New Brunswick. 306 pp.

Swanson, G. A., and M. I. Meyer. 1977. Impact of fluctuating water levels on feeding ecology of breeding blue-winged teal. J. Wildl. Mgmt. 41:426-433.

Swindale, D. N., and J. T. Curtis. 1957. Phytosociology of the larger submerged plants in Wisconsin lakes. Ecol. 38:397–407.

Teal, J., and M. Teal. 1969. Life and death of the salt marsh. Audubon/Ballantine Books, New York. 274 pp.

Threinen, C. W., and W. T. Helm. 1954. Experiments and observations designed to show carp destruction of aquatic vegetation. J. Wildl. Mgmt. 18:247-251.

Toburen, C. 1974. Reclaiming Boulder County gravel pit as a wildlife area. Boulder Valley Soil Cons. Dist., Boulder, Colorado. 41 pp.

Trauger, D. L., and J. H. Stoudt. 1978. Trends in waterfowl populations and habitats on study areas in Canadian parklands. Trans. N. Am. Wildl. Nat. Resour. Conf. 43:187-205.

Tryon, C. A., Jr. 1954. The effect of carp exclosures on growth of submerged aquatic vegetation in Pymatuning Lake, Pennsylvania. J. Wildl. Mgmt. 18:251-254.

U.S. Soil Conservation Service. 1991. Hydric soils of the United States. 3rd ed. GPO, Washington, D.C. 523–416/40289.

U.S. Soil Conservation Service. 1992. Wetland restoration, enhancement, or creation. Eng. Field Handbk. 13. U.S. Dept. Agric., Soil Cons. Serv., Washington, D.C. 79 pp.

U.S. Fish and Wildlife Service and Canadian Wildlife Service. 1992. Status of waterfowl and fall flight forecast. U.S. Fish & Wild. Serv. Off. of Migratory Bird Mgmt., Laurel, Maryland. 30 pp.

U.S. Fish and Wildlife Service and U.S. Department of Commerce. 1993. National survey of fishing, hunting and wildlife-associated recreation. GPO, Washington, D.C. 124 pp. plus appendices.

Utschick, H. 1976. Die Wasservogel als Indikatoren für den ökologischen Zustand von Seen. Verh. orn. Ges. Bayer 22:395438.

van der Valk, A. G. 1981. Succession in wetlands: a Gleasonian approach. Ecol. 62:688–696.

van der Valk, A. G., and C. B. Davis. 1978. The role of seed banks in the vegetation dynamics of prairie glacial marshes. Ecol. 59:322-335.

van der Valk, A. G., C. B. Davis, J. L. Baker, and C. E. Beer. 1979. Natural fresh water wetlands as nitrogen and phosphorus traps for land runoff. Pp. 457–467 *in* Greeson, P. E., J. R. Clark, and J. E. Clark (eds.), Wetland functions and values: the state of our understanding. Amer. Water Resour. Assoc., Minneapolis, Minnesota. 674 pp.

Vermeer, K. 1968. Ecological aspects of ducks nesting in high densities among larids. Wilson Bull. 80:78-83.

Verry, E. S. 1985. Selection of water impoundment sites in the Lake States. Pp. 31-38 *in* Knighton, M.D. (ed.), Water impoundments for wildlife; a habitat management workshop. U.S. For. Serv. N. Cent. For. Exp. Sta. Tech. Rep. NC-100, St. Paul, Minnesota. 136 pp.

Voigts, D. K. 1973. Food niche overlap of two Iowa marsh icterids. Condor 75:392-399.

Voigts, D. K. 1976. Aquatic invertebrate abundance in relation to changing marsh vegetation. Am. Midl. Nat. 95:312-322.

Ward, E. 1942. Phragmites management. Trans. N. Am. Wildl. Conf. 7:294-298.

Ward, H. B., and G. C. Whipple. 1959. Freshwater biology. W. Edmondson (ed.). John Wiley and Sons, New York. 1,248 pp.

Weisner, S. E. B. 1991. Within-lake patterns in depth penetration of emergent vegetation. Freshwater Biol. 26:133–142.

Weller, M. W. 1961. Breeding biology of the least bittern. Wilson Bull.73:11–35.

Weller, M. W. 1972. Ecological studies of Falkland Islands' waterfowl. Wildfowl 23:25–44.

Weller, M. W. 1975a. Studies of cattail in relation to management for marsh wildlife. Iowa St. J. Sci. 49:383–412.

Weller, M. W. 1975b. Ecology and behaviour of the South Georgia pintail *Anas g. georgica*. Ibis 117:217–231.

Weller, M. W. 1978a. Management of freshwater marshes for wildlife. Pp. 267-284 *in* Good, R. E., D. F. Whigham, and R. L. Simpson (eds.), Freshwater wetlands, ecological processes and management potential. Academic Press, New York. 378 pp.

Weller, M. W. 1978b. Density and habitat relationships of blue-winged teal nesting in northwest Iowa. J. Wildl. Mgmt. 43:367-374.

Weller, M. W. 1979. Birds of some Iowa wetlands in relation to concepts of faunal preservation. Proc. Iowa Acad. Sci. 86:81-88.

Weller, M. W. 1989a. Plant and water-level dynamics in an east Texas shrub/hardwood bottomland wetland. Wetlands 9:73–88.

Weller, M. W. 1989b. Waterfowl management techniques for wetland enhancement, restoration and creation useful in mitigation procedures. Pp. 105–116 *in* Kusler, J., and M. Kentula (eds.), Wetland creation and restoration, Vol. 2. U.S. Environmental Protection Agency, Washington, D.C.

Weller, M. W., and L. H. Fredrickson. 1974. Avian ecology of a managed glacial marsh. The Living Bird 12:269-291.

Weller, M. W., G. W. Kaufmann, P. A. Vohs, Jr. 1991. Evaluation of wetland development and waterbird response at Elk Creek Wildlife Management Area, Lake Mills, Iowa, 1961 to 1990. Wetlands 11:245–262.

Weller, M. W., and C. E. Spatcher. 1965. Role of habitat in the distribution and abundance of marsh birds. Iowa St. Univ. Agric. & Home Econ. Exp. Sta. Spec. Rep. No. 43. 31 pp.

Weller, M. W., B. H. Wingfield, and J. B. Low. 1958. Effects of habitat deterioration on bird populations of a small Utah marsh. Condor 60:220-226.

Wetzel, R. G. 1975. Limnology. W. B. Saunders Co., Philadelphia, Pennsylvania. 658 pp.

Whyte, R. J., and N. J. Silvy. 1981. Effects of cattle on duck food plants in southern Texas. J. Wildl. Mgmt. 45:512-515.

Winter, T. C. 1989. Hydrologic studies of wetlands in the northern prairie. Pp. 16–54 *in* van der Valk, A. G. (ed.), Northern prairie wetlands. Iowa St. Univ. Press, Ames. 400 pp.

Zagata, M. D. 1985. Mitigation banking by "credits": a Louisiana pilot project. Trans. N. Am. Wildl. Nat. Resour. Conf. 50:475–485.

Index

Milton W. Weller currently is professor, Kleberg Chair in Wildlife Ecology, at Texas A&M University. Formerly he was professor-in-charge of the Fisheries and Wildlife Section at Iowa State University and head of the Department of Entomology, Fisheries, and Wildlife at the University of Minnesota. He has studied wetlands and their associated wildlife in many parts of the world. He is author of *The Island Waterfowl*, editor of *Waterfowl in Winter,* and has served as associate editor of the *Journal of Wildlife Management* and of *Wetlands*.